A HOLE IN SPACE

Also by Larry Niven

PROTECTOR
THE FLIGHT OF THE HORSE

Larry Niven

A Hole in Space

Futura Publications Limited
An Orbit Book

An Orbit Book

First published in Great Britain in 1975
by Futura Publications Limited
49 Poland Street, London W1A 2LG

Copyright © Larry Niven 1974

ISBN 0 8600 78531
Printed in Great Britain by
C. Nicholls & Company Ltd
The Philips Park Press
Manchester.

Futura Publications Ltd
49 Poland Street
LONDON W1A 2LG

ACKNOWLEDGMENTS

Parts of this book were previously published:

"Rammer", *Galaxy Magazine*. Copyright © 1971 by UPD Publishing Corporation.

"The Alibi Machine," *Vertex*. Copyright © 1973 by Mankind Publishing Company.

"A Kind of Murder," *Analog*. Copyright © 1974 by The Condé Nast Publications, Inc. All rights reserved.

"All the Bridges Rusting," *Vertex*. Copyright © 1973 by Mankind Publishing Company.

"There is a Tide," *Galaxy Magazine*. Copyright © 1968 by Galaxy Publishing Corporation.

"Bigger Than Worlds," *Analog*. Copyright © 1974 by The Conde Nast Publications, Inc. All rights reserved.

"$16,940.00," *Alfred Hitchcock's Mystery Magazine*. Copyright © H.S.D. Publications Inc., 1974. All rights reserved.

"The Hole Man," *Analog*. Copyright © 1973 by the Condé Nast Publications Inc. All rights reserved.

"The Fourth Profession," Quark. Copyright © 1971 by Coronet Communications, Inc. All rights reserved.

I started writing ten years ago. I wrote for a solid year and collected nothing but rejection slips.

Most beginning writers can't afford to do that. They take an honest job and write in their spare time, and it takes them five years to make their mistakes, instead of one. Me, I lived off a trust fund.

The trust fund was there because my great-grandfather once made a lot of money in oil. He left behind him a large family of nice people, and we all owe him.

To Edward Lawrence Doheny

Contents

Rammer   1

The Alibi Machine   29

The Last Days of the
    Permanent Floating Riot Club   41

A *Kind* of Murder   53

All the Bridges Rusting   71

There Is a Tide   95

Bigger Than Worlds   111

$16,940.00   127

The Hole Man   131

The Fourth Profession   147

A
Hole
in
Space

Rammer

I

Once there was a dead man.

He had been waiting for two hundred years inside a coffin whose outer shell held liquid nitrogen. There were frozen clumps of cancer all through his frozen body. He had had it bad.

He was waiting for medical science to find him a cure.

He waited in vain. Most varieties of cancer could be cured now, but no cure existed for the billions of cell walls ruptured by expanding crystals of ice. He had known the risk when he took it and had gambled anyway. Why not? He'd been *dying*.

The vaults held millions of frozen bodies. Why not? They'd been *dying*.

Later there was a criminal. His name is forgotten and his crime is a secret, but it must have been a terrible one. The State wiped his personality for it.

Afterward he was a dead man: still warm, still breathing, even reasonably healthy—but empty.

The State had use for an empty man.

Corbett woke on a hard table, aching as if he had slept too long in one position. He started incuriously at a white ceiling.

1

Memories floated back to him of a double-walled coffin and sleep and pain.

The pain was gone.

He sat up at once.

And flapped his arms wildly for balance. Everything felt wrong. His arms would not swing right. His body was too light. His head bobbed strangely on a thin neck. He reached frantically for the nearest support, which turned out to be a blond young man in a white jumpsuit. Corbett missed—his arms were shorter than he had expected. He toppled to his side, shook his head and sat up more carefully.

His arms. Scrawny, knobby—and not his.

The man in the jumpsuit asked, "Are you all right?"

"Yah," said Corbett. His throat was rusty, but that was all right. His new body didn't fit, but it didn't seem to have cancer, either. "What's the date? How long has it been?"

A quick recovery. The checker gave him a plus. "Twenty-one ninety, your dating. You won't have to worry about our dating."

That sounded ominous. Cautiously Corbett postponed the obvious question: What's happened to me? and asked instead, "Why not?"

"You won't be joining our society."

"No? What, then?"

"Several professions are open to you—a limited choice. If you don't qualify for any of them we'll try someone else."

Corbett sat on the edge of the hard operating table. His body seemed younger, more limber, definitely thinner. He was acutely aware that his abdomen did not hurt no matter how he moved.

He asked, "And what happens to me?"

"I've never learned how to answer that question. Call it a matter of metaphysics," said the checker. "Let me detail what's happened to you so far and then you can decide for yourself."

There was an empty man. Still breathing and as healthy as most of society in the year twenty-one ninety. But empty. The electrical patterns in the brain, the worn paths of nervous reflexes; the memories, the personality of the man had all been wiped away.

And there was this frozen thing.

"Your newstapers called you people corpsicles," said the blond man. "I never understood what the tapes meant by that."

"It comes from popsicle. Frozen sherbet." Corbett had used the word himself before he had become one of them. One of the corpsicles, frozen dead.

Frozen within a corpsicle's frozen brain were electrical patterns that could be recorded. The process would warm the brain and destroy most of the patterns, but that hardly mattered, because other things must be done too.

Personality was not all in the brain. Memory RNA was concentrated in the brain, but it ran all through the nerves and the blood. In Corbett's case the clumps of cancer had to be cut away—then the RNA could be extracted from what was left. The operation would have left nothing like a human being. More like bloody mush, Corbett gathered.

"What's been done to you is not the kind of thing we can do twice," said the checker. "You get one chance and this is it. If you don't work out we'll terminate and try someone else. The vaults are full of corpsicles."

"You mean you'd wipe my personality," Corbett said unsteadily. "But I haven't committed a crime. Don't I have any rights?"

The checker looked stunned. Then he laughed. "I thought I'd explained. The man you think you are is dead. Corbett's will was probated long ago. His widow—"

"Damn it, I left money to myself! A trust fund!"

"No good." Though the man still smiled, his face was impersonal, remote, unreachable. A vet smiles reassuringly at a cat due to be fixed. "A dead man can't own property—that was settled in the courts long ago. It wasn't fair to the heirs. It took the money out of circulation."

Corbett jerked an unexpectedly bony thumb at his bony chest. "But I'm alive now."

"Not in law. You can earn your new life; the State will give you a new birth certificate and citizenship if you give the State good reason."

Corbett sat for a moment, absorbing that. Then he got off the table. "Let's get started then. What do you need to know about me?"

"Your name."

"Jerome Corbett."

"Call me Pierce." The checker did not offer to shake hands. Neither did Corbett, perhaps because he sensed the man would not respond, perhaps because they were both noticeably overdue for a bath. "I'm your checker. Do you like people? I'm just asking. We'll test you in detail later."

"I get along with the people around me but I like my privacy."

The checker frowned. "That narrows it more than you might think. This isolationism you called privacy was, well, a passing fad. We don't have the room for it—or the inclination either. We can't sent you to a colony world . . ."

"I might make a good colonist."

"You'd make terrible breeding stock. Remember, the genes aren't yours. No. You get one choice, Corbett. Rammer."

"Rammer?"

" 'Fraid so."

"That's the first strange word you've used since I woke up. In fact—hasn't the language changed at all? You don't even have an accent."

"Part of the job. I learned your speech through RNA training. You'll learn your trade the same way if you get that far. You'll be amazed how fast you can learn with RNA shots to help you along. But you'd better be right about liking your privacy, Corbett. Can you take orders?"

"I was in the army."

"What does that mean?"

"Yes."

"Good. Do you like strange places and faraway people—or vice versa?"

"Both." Corbett smiled hopefully. "I've raised buildings all over the world. Can the world use another architect?"

"No. Do you feel that the State owes you something?"

There could be but one answer to that. "No."

"But you had yourself frozen. You must have felt that the future owed you something."

"Not at all. It was a good risk. I was dying."

"Ah." The checker looked him over thoughtfully. "If you had something to believe in, perhaps dying wouldn't mean so much."

Corbett said nothing.

They gave him a short word-association test in English. The test made Corbett suspect that a good many corpsicles must date from near his own death. They took a blood sample, then exercised Corbett to exhaustion on a treadmill and took another blood sample. They tested his pain threshhold by direct nerve stimulation—excruciatingly unpleasant—and took another blood sample. They gave him a Chinese puzzle and told him to take it apart.

Pierce then informed him that the testing was over.

"After all, we already know the state of your health."

"Then why the blood samples?"

The checker looked at him for a moment. "You tell me."

Something about that look gave Corbett the creepy feeling that he was on trial for his life. The feeling might have been caused only by the checker's rather narrow features, his icy blue gaze and abstracted smile. Still . . . Pierce had stayed with him all through the testing, watching him as if Corbett's behavior were a reflection on Pierce's judgment. Corbett thought carefully before he spoke.

"You have to know how far I'll go before I quit. You can analyze the blood samples for adrenalin and fatigue poisons to find out just how much I was hurting, just how tired I really was."

"That's right," said the checker.

Corbett had survived again.

He would have given up much earlier on the pain test. But at some point Pierce had mentioned that Corbett was the fourth corpsicle personality to be tested in that empty body.

He remembered going to sleep that last time, two hundred years ago.

His family and friends had been all around him, acting like mourners. He had chosen the coffin, paid for vault space, and made out his Last Will and Testament, but he had not thought of it as *death*. He had been given a shot. The eternal pain had drifted away in a soft haze. He had gone to sleep.

He had drifted off wondering about the future, wondering what he would wake to. A vault into the unknown. World government? Interplanetary spacecraft? Clean fusion power? Strange clothing, body paints, nudism?

Or crowding, poverty, all the fuels used up, power provided by cheap labor? He had thought of those, but it was all right. They would not be able to afford to wake him if they were that poor. The world he dreamed of in those last moments was a rich world, able to support such luxuries as Jerome Corbett.

It looked as if he weren't going to see too damn much of it.

A guard led Corbett away after the testing. He walked with a meaty hand wrapped around Corbett's thin upper arm. Leg irons would have been no more effective had Corbett thought of escaping. The guard took him up a narrow plastic staircase to the roof.

The noon sun blazed in a blue sky that shaded to yellow, then brown at the horizon. Green plants grew in close-packed rows on parts of the roof. Elsewhere many sheets of something glassy were exposed to the sunlight.

Corbett caught one glimpse of the world from a bridge between two roofs. It was a cityscape of close-packed buildings, all of the same cold cubistic design.

And Corbett was impossibly high on a narrow strip of concrete with no guardrails at all. He froze. He stopped breathing.

The guard did not speak. He tugged at Corbett's arm, not hard, and watched to see what he would do. Corbett pulled himself together and went on.

The room was all bunks: two walls of bunks with a gap between. The light was cool and artificial, but outside it was nearly noon. Could they be expecting him to sleep?

The room was big, a thousand bunks big. Most of the bunks were full. A few occupants watched incuriously as the guard showed Corbett which bunk was his. It was the bottommost in a stack of six. Corbett had to drop to his knees and roll to get into it. The bedclothes were strange; silky and very smooth, even slippery—the only touch of luxury in that place. But there was no top sheet, nothing to cover him. He lay on his side, looking out at the dormitory from near floor level.

Three things were shocking about that place.

One was the smell. Apparently perfumes and deodorants had been another passing fad. Pierce had been overdue for a bath. So was Corbett's new self. Here the smell was rich.

The second was the double bunks, four of them in a vertical stack, wider than the singles and with thicker mattresses. The doubles were for loving, not sleeping. What shocked Corbett was that they were right out in the open, not hidden by so much as a gauze curtain.

The same was true of the toilets.

How can they live like this?

Corbett rubbed his nose and jumped—and cursed at himself for jumping. It was the third time he had done that. His own nose had been big and fleshy and somewhat shapeless. But the nose he now rubbed automatically when trying to think was small and narrow with a straight, sharp edge. He might very well get used to the smell and everything else before he got used to his own nose.

Some time after dusk a man came for him. A broad, brawny type wearing a gray jumper and a broad expressionless face, the guard was not one to waste words. He found Corbett's bunk, pulled Corbett out by one arm and led him stumbling away. Corbett was facing Pierce before he was fully awake.

In annoyance he asked, "Doesn't anyone else speak English?"

"No," said the checker.

Pierce and the guard guided Corbett to a comfortable armchair facing a wide curved screen. They put padded earphones on him. They set a plastic bottle of clear fluid on a shelf over his head. Corbett noticed a clear plastic tube tipped with a hypodermic needle.

"Breakfast?"

Pierce missed the sarcasm. "You'll have one meal each day—after learning period and exercise." He inserted the hypodermic into a vein in Corbett's arm. He covered the wound with a blob of what might have been silly putty.

Corbett watched it all without emotion. If he had ever been afraid of needles the months of pain and cancer had worked it out of him. A needle was surcease, freedom from pain for a time.

"Learn now," said Pierce. "This knob controls speed. The volume is set for your hearing. You may replay any section once. Don't worry about your arm—you can't pull the tube loose."

"There's something I wanted to ask you, only I couldn't remember the word. What's a rammer?"

"Starship pilot."

Corbett studied the checker's face. "You're kidding."

"No. Learn now." The checker turned on Corbett's screen and went away.

II

A rammer was the pilot of a starship.

The starships were Bussard ramjets. They caught interstellar hydrogen in immaterial nets of electromagnetic force, guided and compressed and burned the hydrogen for thrust. Potentially there was no limit at all on their speed. They were enormously powerful, enormously complex, enormously expensive.

Corbett found it incredible that the State would trust so much value, such devastating power and mass to one man. To a man two centuries dead! Why, Corbett was an architect, not an astronaut. It was news to him that the concept of the Bussard ramjet predated his own death. He had watched the Apollo XI and XIII flights on television and that had been the extent of his interest in spaceflight until now.

Now his life depended on his "rammer" career. He never doubted it. That was what kept Corbett in front of the screen with the earphones on his head for fourteen hours that first day. He was afraid he might be tested.

He didn't understand all he was supposed to learn. But he was not tested either.

The second day he began to get interested. By the third day he was fascinated. Things he had never understood—relativity and magnetic theory and abstract mathematics—he now grasped intuitively. It was marvelous!

And he ceased to wonder why the State had chosen Jerome Corbett. It was always done this way. It made sense, all kinds of sense.

The payload of a starship was small and its operating lifetime was more than a man's lifetime. A reasonably safe life-support system for one man occupied an unreasonably high proportion of the payload. The rest must go for biological package probes.

As for sending a citizen, a loyal member of the State—what for? The times would change enormously before a starship could return. The State itself might change beyond recognition. A returning rammer must adjust to a whole new culture—with no way of telling in advance what it might be like.

Why not pick a man who had already chosen to adjust to a new culture? A man whose own culture was two centuries dead before the trip started?

And a man who already owed the State his life?

The RNA was most effective. Corbett stopped wondering about Pierce's dispassionately possessive attitude. He began to think of himself as property being programmed for a purpose.

And he learned. He skimmed microtaped texts as if they were already familiar. The process was heady. He became convinced that he could rebuild a ramship with his bare hands, given the parts. He had loved figures all his life, but abstract mathematics had been beyond him until now. Field theory, monopole field equations, circuitry design. When to suspect the presence of a gravitational point source—how to locate it, use it, avoid it.

The teaching chair was his life. The rest of his time—exercise, dinner, sleep—seemed vague, uninteresting.

He exercised with about twenty others in a room too small for the purpose. Like Corbett, the others were lean and stringy, in sharp contrast to the brawny wedge-shaped men who were their guards. They followed the lead of a guard, running in place because there was no room for real running, forming precise rows for scissors jumps, pushups, situps.

After fourteen hours in a teaching chair Corbett usually enjoyed the jumping about. He followed orders. And he wondered about the stick in a holster at the guard's waist. It looked like a cop's baton. It might have been just that—except for the hole in one end. Corbett never tried to find out.

Sometimes he saw Pierce during the exercise periods. Pierce and the men who tended the teaching chairs were of a third type: well fed, in adequate condition, but just on the verge of being overweight. Corbett thought of them as Olde American types.

From Pierce he learned something of the other professions open to a revived corpsicle/reprogramed criminal. Stoop labor; intensive hand cultivation of crops. Body servants. Handicrafts. And easily taught repetitive work. And the hours! The corpsicles were expected to work fourteen hours a day. And the crowding!

He was leading that life now. Fourteen hours to study, an hour of heavy exercise, an hour to eat and eight hours to sleep in a dorm that was two solid walls of people.

"Time to work, time to eat, time to sleep! Elbow to elbow every minute! The poor bastards," he said to Pierce. "What kind of a life is that?"

"It lets them repay their debt to the State as quickly as possible. Be reasonable, Corbett. What would a corpsicle do with his off hours? He has no social life—he has to learn one by observing citizens. Many forms of corpsicle labor involve proximity to citizens."

"So they can look up at their betters while they work? That's no way to learn. It would take ... I get the feeling we're talking about *decades* of this kind of thing."

"Thirty years' labor generally earns a man his citizenship. That gets him a right to work—which then gets him a guaranteed base income he can use to buy education tapes and shots. And the medical benefits are impressive. We live longer than you used to, Corbett."

"Meanwhile it's slave labor. Anyway, none of this applies to me—"

"No, of course not. Corbett, you're wrong to call it slave labor. A slave can't quit. You can change jobs any time you like. There's a clear freedom of choice."

Corbett shivered. "Any slave can commit suicide."

"Suicide, my ass," the checker said distinctly. If he had any-thing that could be called an accent it lay in the precision of his pronunciation. "Jerome Corbett is dead. I could have given you his intact skeleton for a souvenir."

"I don't doubt it." Corbett saw himself tenderly polishing his own white bones. But where could he have kept such a thing?

"Well, then. You're a brain-wiped criminal, justly brain-wiped, I might add. Your crime has cost you your citizenship, but you still have the right to change professions. You need only ask for another personality. What slave can change jobs at will?"

"It would feel like dying."

"Nonsense. You go to sleep, that's all. When you wake up you've got a different set of memories."

The subject was an unpleasant one. Corbett avoided it from then on. But he could not avoid talking to the checker. Pierce was the only man in the world he could talk to. On the days Pierce failed to show up he felt angry, frustrated.

Once he asked about gravitational point sources. "My time didn't know about those."

"Yes, it did. Neutron stars. You had a number of pulsars lo-cated by nineteen seventy, and the math to describe how a pulsar decays. The thing to watch for is a decayed pulsar di-rectly in your path."

"Oh."

Pierce regarded him in some amusement. "You really don't know much about your own time, do you?"

"Astrophysics wasn't my field. And we didn't have your learning techniques." Which reminded him of something. "Pierce, you said you learned English with RNA injections. Where did the RNA come from?"

Pierce grinned and left.

Corbett did not want to die. He was utterly, disgustingly healthy and twenty years younger than he had been at death. He found his rammer education continually fascinating. If only they would stop treating him like property . . .

Corbett had been in the army, but that had been twenty years before his death. He had learned to take orders, but

never to like it. What had galled him then had been the basic
assumption of his inferiority. But no army officer in Corbett's
experience had believed in Corbett's inferiority as completely
as did Pierce and Pierce's guards.

The checker never repeated a command, never seemed even
to consider that Corbett would refuse. If Corbett refused,
once, he knew what would happen. And Pierce knew that he
knew. No army could have survived in such a state. The attitude better fitted a death camp.

They must think I'm a zombie....

Corbett carefully did not pursue the thought. He was a
corpse brought back to life—but not all the way.

The life was not pleasant. His last-class citizenship was gall-
ing. There was nobody to talk to—nobody but Pierce, whom
he was learning to hate. He was hungry most of the time. The
single daily meal barely filled his belly and it would not stay
full. No wonder he had wakened so lean.

More and more he lived in the teaching chair. Vicariously
he became a rammer then and the impotence of his life was
changed to omnipotence. Starman! Riding the fire that feeds
the suns, scooping fuel from interstellar space itself, spreading
electromagnetic fields like wings hundreds of miles out ...

Two weeks after the State had wakened him from the dead,
Corbett was given his course.

He relaxed in a chair that was not quite a contour couch.
RNA solution dripped into him. The needle no longer bothered him—he never noticed it. The teaching screen held a
map of his course, in green lines in three-space. Corbett had
stopped wondering how the three-dimensional effect was
achieved.

The scale was shrinking as he watched.

Two tiny blobs, and a glowing ball surrounded by a faintly
glowing corona. This part of his course he already knew. A
linear accelerator would launch him from the moon, boost
him to Bussard ramjet speeds and hurl him at the sun. Solar
gravity would increase his speed while his electromagnetic
fields caught and burned the solar wind itself. Then out, still
accelerating, to the stars ...

In the teaching screen the scale shrank horrendously. The

distances between stars were awesome, terrifying. Van Maanan's Star was twelve light-years away.

He would begin deceleration a bit past the midpoint. The matching would be tricky. He must slow enough to release the biological package probe—but not enough to drop him below ram speeds. In addition he must use the mass of the star for a course change. There was no room for error here.

Then on to the next target, which was even farther away. Corbett watched . . . and he absorbed . . . and a part of him seemed to have known everything all along even while another part was gasping at the distances. Ten stars, all yellow dwarves of the Sol type, an average of fifteen light-years apart—though he would cross one gap of fifty-two light-years. He would almost touch lightspeed on that one. Oddly enough, the Bussard ramjet effect would improve at such speeds. He could take advantage of the greater hydrogen flux to pull the fields closer to the ship, to intensify them.

Ten stars in a closed path, a badly bent and battered ring leading him back to the solar system and Earth. He would benefit from the time he spent near the speed of light. Three hundred years would pass on Earth, but Corbett would only live through two hundred years of ship's time—which implied some kind of suspended animation technique.

It didn't hit him the first time through, or the second; but repetition had been built into the teaching program. It didn't hit him until he was on his way to the exercise room.

Three hundred years?

Three hundred years!

III

It wasn't night, not really. Outside it must be midafternoon. Indoors, the dorm was always coolly lit, barely brightly enough to read if there had been any books. There were no windows.

Corbett should have been asleep. He suffered every minute he spent gazing out into the dorm. Most of the others were asleep, but a couple made noisy love on one of the loving bunks. A few men lay on their backs with their eyes open and two women talked in low voices. Corbett didn't know the language. He had been unable to find anyone who spoke English.

He suspected that there were two shifts, that someone slept in his bunk, mornings—but he could prove nothing. The slippery sheets must be fantastically easy to clean. Just hose them down.

Corbett was desperately homesick.

The first few days had been the worst.

He had stopped noticing the smell. If something reminded him he could sniff the traces of billions of human beings. Otherwise the odor was part of the environment.

But the loving bunks bothered him. When they were in use he watched. When he forced himself not to watch he listened. He couldn't help himself. But he had turned down two sign-language invitations from a small brunette with straggly hair and a pretty, elfin face. Make love in public? He couldn't.

He could avoid using the loving bunks, but not the exposed toilets. That was embarrassing. The first time he was able to force himself only by staring rigidly at his feet. When he pulled on his jumper and looked up a number of sleepers were watching him in obvious amusement. The reason might have been his self-consciousness or the way he dropped his jumper around his ankles, or he may have been out of line. A pecking order determined who might use the toilets before whom. He still hadn't figured out the details.

Corbett wanted to go home.

The idea was unreasonable. His home was gone and he would have gone with it without the corpsicle crypts. But reason was of no use in this instance—he wanted to go home. Home to Mirian, who long since must have died of old age. Home to anywhere: Rome, San Francisco, Kansas City, Hawaii, Brasilia—he had lived in all those places, all different, but all home. Corbett had been a born traveler, "at home" anywhere—but he was not at home here and never would be.

Now they would take here away from him. Even this world of four rooms and two roofs—this world of elbow-to-elbow mutes and utter slavery, this world of which he knew nothing—would have vanished when he returned from the stars.

Corbett rolled over and buried his face in his arms. If he didn't sleep he would be groggy tomorrow. He might miss something essential. They had never tested his training. Read that: Not yet, not yet ...

He dozed.

He came awake suddenly, already up on one elbow, groping for some elusive thought.

Ah.

Why haven't I been wondering about the biological package probes?

A moment later he did wonder.

What are the biological package probes?

But the wonder was that he had never wondered.

He knew what and where they were: heavy fat cylinders arranged around the waist of the starship's hull. Ten of these, each weighing almost as much as Corbett's own life-support system. He knew their mass distribution. He knew the clamp system that held them to the hull and he could operate and repair the clamps under various extremes of damage. He almost knew where the probes went when released; it was just on the tip of his tongue—which meant he had had the RNA shot but had not yet seen the instructions.

But he did not know what the probes were for.

It was like that with the ship, he realized. He knew everything there was to know about a seeder ramship, but nothing at all about the other kinds of ramship or interplanetary travel or ground-to-orbit vehicles. He knew that he would be launched by linear accelerator from the moon. He knew the design of the accelerator—he could see it, three hundred and fifty kilometers of rings standing on end in a line across a level lunar mare. He knew what to do if anything went wrong during launch. And that was all he knew about the moon and lunar installations and lunar conquest, barring what he had watched on television two hundred years ago.

What was going on out there? In the two weeks since his arrival (awakening? resuscitation?) he had seen four rooms and two rooftops, glimpsed a fantastic cityscape from a bridge and talked to one man who was not interested in telling him anything. What had happened in two hundred years?

These men and women who slept around him. Who were they? Why were they here? He didn't even know if they were corpsicles or contemporary. Probably contemporary. Not one of them was self-conscious about the facilities.

Corbett had raised his buildings in all sorts of strange places, but he had never jumped blind. He had always brushed up on the language and studied the customs before he went. Here he had no handle, nowhere to start. He was lost.

If only he had someone he could really talk to!

He was learning in enormous gulps, taking in volumes of knowledge so broad that he hadn't realized how rigidly bounded they were. The State was teaching him only what he needed to know or might need to know some time. Every bit of information was aimed straight at his profession.

Rammer.

He could see the reasoning. He would be gone for several centuries. Why should the State teach him anything at all about today's technology, customs, geography? There would be trouble enough when he came back if he——come to think of it, who had taught him to call the government the State? He knew nothing of its power and extent. How had he come to think of the State as all-powerful?

It must be the RNA training. With data came attitudes below the conscious level, where he couldn't get at them.

What were they doing to him?

He had lost his world. He would lose this one. According to Pierce, he had lost himself four times already. A condemned criminal had had his personality wiped four times. Now Corbett's beliefs and motivations were being lost bit by bit to the RNA solution as the State made him over into a rammer.

Was there nothing that was his?

He failed to see Pierce at exercise period. It was just as well. He was somewhat groggy. As usual, he ate dinner like a starving man. He returned to the dorm, rolled into his bunk and was instantly asleep.

He looked up during study period the next day and found Pierce watching him. He blinked, fighting free of a mass of data on the attitude jet system that bled plasma from the inboard fusion plant that was also the emergency electrical power source—and asked, "Pierce, what's a biological package probe?"

"I would have thought they would teach you that. You know what to do with the probes, don't you?"

"The teaching widget gave me the procedure two days ago. Slow up for certain systems, kill the fields, turn a probe loose and speed up again."

"You don't have to aim them?"

"No, I guess they aim themselves. But I have to get them down to a certain relative velocity to get them into the system."

"Amazing. They must do all the rest of it themselves." Pierce shook his head. "I wouldn't have believed it. Well, Corbett, the probes steer for a terrestrial world with a reducing atmosphere. They outnumber oxygen-nitrogen worlds about three to one in this arm of the galaxy and probably everywhere else, too—as you may know, if your age got that far."

"But what do the probes do?"

"They're biological packages. Bacteria. The idea is to turn a reducing atmosphere into an oxygen atmosphere just the way certain bacteria did it for Earth, something like fifteen-times-ten-to-the-eighth years ago." The checker smiled—barely. His small narrow mouth wasn't built to express any great emotion, "You're part of a big project, Corbett."

"Good Lord. How long does it take?"

"We think about fifty thousand years. Obviously we've never had a chance to measure it."

"But, good Lord! Do you really expect the State to last that long? Does even the State expect to last that long?"

"That's not your affair, Corbett. Still—" Pierce considered. "—I don't suppose I do. Or the State does. But humanity will last. One day there will be men on those worlds. It's a Cause, Corbett. The immortality of the species. A thing bigger than one man's life. And you're part of it."

He looked at Corbett expectantly.

Corbett was deep in thought. He was running a finger tip back and forth along the straight line of his nose.

Presently he asked, "What's it like out there?"

"The stars? You're—"

"No, no, no. The city. I catch just a glimpse of it twice a day; cubistic buildings with elaborate carvings at the street level—"

"What the bleep is this, Corbett? You don't need to know anything about Selerdor. By the time you come home the whole city will be changed."

"I know, I know. That's why I hate to leave without seeing something of this world. I could be going out to die—"

Corbett stopped. He had seen that considering look before, but he had never seen Pierce actually angry.

The checker's voice was flat, his mouth pinched tight. "You think of yourself as some kind of tourist."

"So would you if you found yourself two hundred years in the future. If you didn't have that much curiosity you wouldn't be human."

"Granted that I'd want to look around. I certainly wouldn't demand it as a right. Corbett, what were you thinking when you foisted yourself off on the future? Did you think the future owed you a debt? It's the other way around—and time you realized it!"

Corbett was silent.

"I'll tell you something. You're a rammer because you're a born tourist. We tested you for that. You like the unfamiliar, it doesn't send you scuttling back to something safe and known. That's rare." The checker's eyes said: And that's why I've decided not to wipe your personality yet. His mouth said, "Was there anything else?"

Corbett pushed his luck. "I'd like a chance to practice with a computer like the ship's computer-autopilot."

"We don't have one. But you'll get your chance in two days. You're leaving then."

IV

The next day he received his instructions for entering the solar system. He was to try anything and everything to make contact, up to and including flashing his attitude jets in binary code. The teaching widget was fanatical on the subject.

He found that he would not be utterly dependent on rescue ships. He could slow the ramship by braking directly into the solar wind until the proton flux was too slow to help him. He could then proceed on attitude jets, using whatever hydrogen was left in the emergency tank. A nearly full tank would actually get him to the moon and land him there.

The State was through with him when he dropped his last probe. It was good of the State to provide for his return, Corbett thought—and then he shook himself. The State was not altruistic. It wanted the ship back.

Now, more than ever, Corbett wanted a chance at the computer-autopilot.

He found one more chance to talk to the checker.

"A three-hundred-year round trip—maybe two hundred, ship's time," said Corbett. "I get some advantage from relativity. But Pierce, you don't really expect me to live two hundred years, do you? With nobody to talk to?"

"The cold sleep treatment—"

"Even so."

Pierce frowned. "You haven't studied medicine. I'm told that cold sleep has a rejuvenating effect over long periods. You'll spend perhaps twenty years awake and the rest in cold sleep. The medical facilities are automatic; I'm sure you've been instructed how to use them. They are adequate. Do you think we'd risk your dying out there between the stars, where it would be impossible to replace you?"

"No."

"Was there anything else you wanted to see me about?"

"Yes." He had decided not to raise the subject. Now he changed his mind. "I'd like to take a woman with me. The life-support system would hold two of us easily enough. I worked it out. We'd need another cold sleep chamber, of course."

For two weeks this had been the only man Corbett could talk to. At first he had found Pierce unfathomable, unreadable, almost inhuman. Since then he had learned to read the checker's face to some extent.

Now he watched Pierce decide whether to terminate Jerome Corbett and start over.

It was a close thing. But the State had spent considerable time and effort on Jerome Corbett. It was worth a try. And so Pierce said, "That would take up some space. You would have to share the rest between you. I do not think you would survive, Corbett."

"But—"

"Look here, Corbett. We know you don't need a woman. If you did you would have taken one by now and we would have wiped you and started over. You've lived in the dormitory for two weeks and you have not used the loving bunks once."

"Damn it, Pierce, do you expect me to make love in public? I can't."

"Exactly."

"But—"

"Corbett, you learned to use the toilet, didn't you? Because you had to. You know what to do with a woman but you are one of those men fortunate enough not to need one. Otherwise you could not be a rammer."

If Corbett had hit the checker then he would have done it knowing that it meant his death. And knowing that, he would have killed Pierce for forcing him to it.

Something like ten seconds elapsed, during which he might have done it. Pierce watched him in frank curiosity.

When he saw Corbett relax he said, "You leave tomorrow, Corbett. Your training is finished. Goodbye."

And Corbett walked out.

The dormitory had been a test. He knew it now. Could he cross a narrow bridge with no handrails? Then he was not pathologically afraid of falling. Could he spend two hundred years alone in the cabin of a starship? Then the silent people around him, five above his head, thousands to either side, must make him markedly uncomfortable. Could he live two hundred years without a woman? Surely he must be impotent.

He returned to the dorm after dinner. They had replaced the bridge with a nearly invisible slab of glass.

Corbett snarled and crossed ahead of the guard. The guard had to hurry to keep up.

He stood between two walls of occupied bunks, looking about him. Then he did a stupid thing.

He had already refrained from killing the checker. He must

have decided to live. What he did, then, was stupid. He knew it.

He looked about him until he found the slender darkhaired girl with the elfin face watching him curiously from near the ceiling. He climbed the rungs between bunks until his face was level with her bunk.

He remembered that the gesture he needed was a quick, formalized one; he didn't know it.

In English he asked, "Come with me?"

She nodded brightly and followed him down the ladder. By then it seemed to Corbett that the dorm was alive with barely audible voices.

The odd one, the rammer trainee.

Certainly a number of the wakeful turned to lie on their sides to watch.

He felt their eyes on the back of his neck as he zipped open his gray jumpsuit and stepped out of it. The dormitory had been a series of tests. At least two of those eyes must belong to someone who would report to Pierce or to Pierce's bosses. But to Corbett they were just like the others, all the eyes curiously watching to see how the speechless one would make out.

And sure enough, he was impotent. It was the eyes—and he was naked. The girl was first concerned, then pitying. She stroked his cheek in apology or sympathy and then she went away and found someone else.

Corbett lay listening to them, gazing at the bunk above him.

He waited for eight hours. Finally a guard came to take him away. By then he didn't care what they did with him.

He didn't start to care until the guard's floating jeep pulled up beneath an enormous .22 long cartridge standing on end. Then he began to wonder. It was too small to be a rocket ship.

But it was one. They strapped him into a contour couch, one of three in a cabin with a single window. There were the guard type and Corbett and a man who might have been Pierce's second cousin once removed. He had the window. He also had the controls.

Corbett's heartbeat quickened. He wondered how it would be.

It was as if he had suddenly become very heavy. He heard no noise except right at the beginning—a sound like landing gear being raised on an airplane. Not a rocket, Corbett thought—and he remembered the tricks a Bussard ramjet could play with magnetic fields. He was heavy and he hadn't slept a wink last night. He went to sleep.

When he awoke he was in free fall. Nobody had tried to tell him anything about free fall. The guard and the pilot watched him curiously to see what he would do.

"Screw you," said Corbett.

It was another test. He got the straps open and pushed himself over to the window. The pilot laughed, caught him and held him while he closed a protective cover over the instruments. Then he let go and Corbett drifted before the window.

His belly was revolving eccentrically. His inner ear was going crazy. His testicles were tight up against his groin and that didn't feel good either. He felt as if the elevator cable had snapped. Corbett snarled within his mind and tried to concentrate on the window. But the Earth was not visible. Neither was the moon. Just a lot of stars, bright enough— quite bright in fact—even more brilliant than they had been above a small boat anchored off Catalina Island one night long ago. He watched them for some time.

Trying to keep his mind off that falling elevator.

He wasn't about to get himself disqualified now.

They ate aboard in free fall. Corbett copied the others, picking chunks of meat and potatoes out of a plastic bag of stew, pulling them through a membrane that sealed itself behind his pick.

"Of all the things I'm going to miss," he told the broad-faced guard, "I'm going to enjoy missing you most. You and your goddamn staring eyes."

The guard smiled placidly and waited to see if Corbett would get sick.

They landed a day after takeoff on a broad plain where the Earth sat nestled in a row of sharp lunar peaks. One day instead of four—the State had expended extra power to get him here. But an Earth-moon flight must be a small thing these days.

The plain was black with blast pits. It must have been a landing field for decades. Enormous transparent bubbles with trees and buildings inside them clustered near the runway end of the linear accelerator, and spacecraft of various types were scattered about the plain.

The biggest was Corbett's ramship: a silver skyscraper lying on its side. The probes were in place, giving the ship a thick-waisted appearance. To Corbett's trained eye it looked ready for takeoff.

Corbett donned his suit first, while the pilot and guard watched to see if he would make a mistake. It was the first time he had seen a suit off the teaching screen. He took it slowly.

There was an electric cart. Apparently Corbett was not ex-pected to know how to walk on an airless world. He thought to head for one of the domes, but the guard steered straight for the ramship. It was a long way off.

It had become unnervingly large when the guard stopped underneath.

The guard said, "Now you inspect your ship."

"You can talk?"

"Yes. Yesterday, a quickie course."

"Oh."

"Three things wrong with your ship. You find all three. You tell me, I tell him."

"Him? Oh, the pilot. Then what?"

"Then you fix one of the things, we fix the others. Then we launch you."

It was another test, of course, maybe the last. Corbett was furious. He started immediately with the field generators and gradually he forgot the guard and the pilot and the sword still hanging over his head. He knew this ship. As it had been with the teaching chair, so it was with the ship itself. Corbett's impotence changed to omnipotence. The power of the beast, the intricacy, the potential, the . . . the hydrogen tank held far too much pressure. That wouldn't wait.

"I'll slurry this now," he told the guard. "Get a tanker over there to top it off." He bled gas slowly through the gauge, lowering the fuel's vapor pressure without letting fuel boil out the gauge itself. When he finished the liquid hydrogen would be slushy with frozen crystals under near-vacuum pressure.

He finished the external inspection without finding anything more. It figured: the banks of dials held vastly more information than a man's eyes could read through opaque titanalloy skin.

The airlock was a triple-door type, not so much to save air as to give him an airlock even if he lost a door somehow. Corbett shut the outer door, used the others as green lights indicated he could. He looked down at the telltales under his chin as he started to unclamp his helmet.

Vacuum?

He stopped. The ship's gauges said air. The suit's said vacuum. Which was right? Come to think of it, he hadn't heard any hissing. Just how soundproof was his helmet?

Just like Pierce to wait and see if he would take off his helmet in vacuum. Well, how to test?

Hah! Corbett found the head, turned on a water spigot. The water splashed oddly in lunar gravity. It did not boil.

Corbett doffed his helmet and continued his inspection.

There was no way to test the electromagnetic motors without causing all kinds of havoc in the linear accelerator. He checked out the telltales as best he could, then concentrated on the life-support mechanisms. The tailored plants in the air system were alive and well. But the urea absorption mechanism was plugged somehow. That would be a dirty job. He postponed it.

Did a flaw in his suit constitute a flaw in the ship?

He decided to finish the inspection. The State might have missed something. It was his ship, his life.

The cold sleep chamber was like a great coffin, a corpsicle coffin. Corbett shuddered at the sight of it. It reminded him of two hundred years spent waiting in liquid nitrogen. He wondered again if Jerome Corbett were really dead—and then he shook off the wonder and went to work.

No flaw there.

The computer was acting vaguely funny.

He had a hell of a time tracing the problem. There was a minute break in one superconducting circuit, so small that some current was leaking through anyway, by inductance. Bastards. He donned his suit and went out to report.

The guard heard him out, consulted with the other man, then told Corbett, "You did good. Now finish with the topping off procedure. We fix the other things."

"There's something wrong with my suit too."

"New suit aboard now."

"I want some time with the computer," said Corbett. "I want to be sure it's all right now."

"We fix it good. When you top off fuel you leave."

That suddenly, Corbett felt a vast sinking sensation. The whole moon was dropping away under him.

They launched him hard. Corbett saw red before his eyes, felt his cheeks dragged far back toward his ears. This ship would be all right. It was built to stand electromagnetic eddy currents from any direction.

He survived. He fumbled out of his couch in time to watch the moonscape flying under him, receding, a magnificent view.

There were days of free fall. He was not yet moving at ramspeeds. But the State had aimed him inside the orbit of Mercury, straight into the thickening solar wind. Protons. Thick fuel for the ramfields and a boost from the sun's gravity.

Meanwhile he had several days. He went to work with the computer.

At one point it occurred to him that the State might monitor his computer work. He shrugged it off. Probably it was too late for the State to stop him now. In any case, he had said too much already.

He finished his work at the computer and got answers that satisfied him. At higher speeds the ram fields were self-reinforcing—they would support themselves and the ship. He could find no upper limit to the velocity of a ramship.

With all the time in the world, then, he sat down at the control console and began to play with the ramfields.

They emerged like invisible wings and he felt the buffeting of badly controlled bursts of fusing hydrogen. He kept the fields close to the ship, fearful of losing the balance here, where the streaming of protons was so uneven. He could feel how he was doing—he could fly this ship by the seat of his pants, with RNA training to help him.

He felt like a giant. This enormous, phallic, germinal fly-ingthing of metal and fire! Carrying the seeds of life for worlds that had never known life, he roared around the sun and out. The thrust dropped a bit then, because he and the solar wind were moving in the same direction. But he was catching it in his nets like wind in a sail, guiding it and burning it and throwing it behind him. The ship moved faster every second.

This feeling of power—enormous masculine power—had to be partly RNA training. At this point he didn't care. Part was him, Jerome Corbett.

Around the orbit of Mars, when he was sure that a glimpse of sunlight would not blind him, he opened all the ports. The sky blazed around him. There were no planets nearby and all he saw of the sky was myriads of brilliant pinpoints, mostly white, some showing traces of color. But there was more to see. Fusing hydrogen made a ghostly ring of light around his ship.

It would grow stronger. So far his thrust was low, somewhat more than enough to balance the thin pull of the sun.

He started his turn around the orbit of Jupiter by adjusting the fields to channel the proton flow to the side. That helped his thrust, but it must have puzzled Pierce and the faceless State. They would assume he was playing with the fields, testing his equipment. Maybe. His curve was gradual—it would take them a while to notice.

This was not according to plan. Originally he had intended to go as far as Van Maanan's Star, then change course. That would have given him $2 \times 15 = 30$ years' head start, in case he was wrong, in case the State could do something to stop him even now. Fifteen years for the light to show them his change in course; fifteen more before retaliation could reach him.

It was wise; but he couldn't do it. Pierce might die in thirty years. Pierce might never know he had failed—and that thought was intolerable.

His thrust dropped to almost nothing in the outer reaches of the system. Protons were thin out here. But there were enough to push his velocity steadily higher and that was what counted. The faster he went, the greater the proton flux. He was on his way.

He was beyond Neptune when the voice of Pierce the checker came to him, saying, "This is Peerssa for the State, Peerssa for the State. Answer, Corbett. Do you have a malfunction? Can we help? We cannot send rescue but we can advise. Peerssa for the State, Peerssa for the State—"

Corbett smiled tightly. Peerssa? The checker's name had changed pronunciation in two hundred years. Pierce had slipped back to an old habit, RNA lessons forgotten. He must be upset about something.

Corbett spent twenty minutes finding the moon base with his signal laser. The beam was too narrow to permit sloppy handling.

When he had it adjusted he said, "This is Corbett for himself, Corbett for himself. I'm fine. How are you?"

He spent more time at the computer. One thing had been bothering him: the return. He planned to be away longer than the State would have expected. Suppose there was nobody on the moon when he returned?

It was a problem, he found. If he could reach the moon on his remaining fuel (no emergencies, remember), he could reach the Earth's atmosphere. The ship was durable; it would stand a meteoric re-entry. But his attitude jets would not land him, properly speaking.

Unless he could cut away part of the ship. The ram field generators would no longer be needed ... Well, he would work it out somehow. Plenty of time. Plenty of time.

The answer took nine hours. "Peerssa for the State. Corbett, we don't understand. You are way off course. Your first target was to be Van Naanan's Star. Instead you seem to be curving around toward Sagittarius. There is no known Earthlike planet in that direction. What the bleep do you think you're doing? Repeating. Peerssa for the State, Peerssa—"

Corbett tried to switch it off. The teaching chair hadn't told him about an off switch. He managed to disconnect a wire. Somewhat later, he located the lunar base with his signal laser and began transmission.

"This is Corbett for himself, Corbett for himself. I'm getting sick and tired of having to find you every damn time I want to say something. So I'll give you this all at once.

"I'm not going to any of the stars on your list.

"It's occurred to me that the relativity equations work better for me the faster I go. If I stop every fifteen light-years to launch a probe, the way you want me to, I could spend two hundred years at it and never get anywhere. Whereas if I just aim the ship in one direction and keep it going, I can build up a ferocious Tau factor.

"It works out that I can reach the galactic hub in twenty-one years, ship's time, if I hold myself down to one gravity acceleration. And, Pierce, I just can't resist the idea. You were the one who called me a born tourist, remember? Well, the stars in the galactic hub aren't like the stars in the arms. And they're packed a quarter to a half light-year apart, according to your own theories. It must be passing strange in there.

"So, I'll go exploring on my own. Maybe I'll find some of your reducing-atmosphere planets and drop the probes there. Maybe I won't. I'll see you in about seventy thousand years, your time. By then your precious State may have withered away, or you'll have colonies on the seeded planets and some of them may have broken loose from you. I'll join one of them. Or—"

Corbett thought it through, rubbing the straight, sharp line of his nose. "I'll have to check it out on the computer," he said. "But if I don't like any of your worlds when I get back, there are always the Clouds of Magellan. I'll bet they aren't more than twenty-five years away, ship's time."

The Alibi Machine

McAllister left the party around eight o'clock.

"Out of tobacco," he told his host apologetically. The police, if they got that far, would discover that that had been a little white lie. There were other parties in Greenwich Village on a Saturday night, and he would be attending one in about, he estimated, twenty minutes.

He took the elevator down. There was a displacement booth in the lobby. He dropped a coin in the slot, smiling fleetingly at himself—he had almost forgotten to take coins—and dialed. A moment later he was outside his own penthouse door in Queens.

He had saved himself the time to let himself in, by leaving his briefcase under a potted plant earlier this evening. He tipped the pot, picked up the briefcase and stepped back into the booth. His conservative paper business suit made him look as if he had just come from work, and the briefcase completed the picture nicely.

He dialed three times. The first number took him to Kennedy International. The second to Los Angeles International. Long distance flicks required the additional equipment available only at what had once been airports: equipment to compensate for the difference in rotational velocity between

29

different points on the Earth. The third number took him to Lucas Anderson's home in the high Sierras.

It was five o'clock here, and the summer sun was still high. McAllister found himself gasping as he left the booth. Why would Anderson want to live at eight thousand feet?

For the view, he supposed; and because Anderson, a freelance writer, did not have to leave his home as often as normal people did. But there was also his love of privacy—and distrust of people.

He rang the bell.

Anderson's look was more surprised than welcoming. "It was tomorrow. After lunch, remember?"

"I know, but—" McAllister hefted the briefcase. "Your royalty accounts arrived this afternoon. A day earlier than we expected. I got to thinking, why not have it out now? Why let you go on thinking you've been cheated a day longer than—"

"Uh huh." Anderson had an imposing scowl. He gave no indication that he was ready to change his mind—and McAllister had nothing to change it with anyway. Publishing companies had always fudged a little on their royalty statements. Sometimes they took a bit too much, and then a writer might rear back on his hind legs and demand an audit.

The difference here was that Brace Books didn't know what McAllister had been doing with Lucas Anderson's accounts.

"Let's just go over these papers," he said with a trace of impatience.

Anderson nodded without enthusiasm, and stepped back, inviting him in.

Did he have company? A glance into the dining nook told McAllister that he did not. A dinner setting for one, laid out with mathematical precision by one or another of Anderson's machines. Anderson's house was a display case of labor saving devices.

How to get him into the living room? But Anderson was leading him there. It was not a big house, and a hostile publisher's assistant would not be invited into the semisacred writing room.

Anderson stopped in the middle of the room. "Spread it on the coffee table."

McAllister circled Anderson as he reached into the

briefcase. His fingers brushed papers, and then the GyroJet, and suddenly his pulse was thundering in his ears. He was afraid.

He'd spent considerable time plotting this. He'd even typed outlines, as for a mystery novel, and burned them afterward. He could produce the royalty statements; they were there in his briefcase, though they would not stand up. Or ... His hand, unseen within the briefcase, clenched into a fist.

He was between Anderson and the picture window when he produced the GyroJet.

The GyroJet: an ancient toy or weapon, depending. It was a rocket pistol, made during the 1960s, then discontinued. This one had been stolen from someone's house and later sold to McAllister, secretly, a full twelve years ago.

A rocket pistol. How could any former Buck Rogers fan have turned down a rocket pistol? He had never shown it to anyone. He had had the thought, even then, that it would be untraceable should he ever want to kill somebody.

The true weapon was the rocket slug. The gun looked like a toy, flimsy aluminum, perforated down the barrel. Anderson might have thought it was a toy—but Anderson was bright. He got the point immediately. He turned to run.

McAllister shot him twice in the back.

He left by the front door. He grinned as he passed the displacement booth. Fifteen years ago there had been people who put their displacement booths inside, in the living room, say. But it made burglaries much too easy.

The alibi machine, the newspapers had called it then. They still did. The advent of the displacement booths had produced one hell of a crime wave. When a man in, say, Hawaii could commit murder in Chicago and be back in the time it would take him to visit the men's room, it did make things a bit difficult for the police. McAllister himself would be at a party in New York ten minutes from now. But first ...

He walked around to the back of the house and stood a moment, looking into the picture window.

He'd thrown a paper tablecloth over Anderson's body. Glass particles on the body would be a giveaway. He'd take the tablecloth with him; and how were the police to know that it was the third bullet, rather than the first, that had shattered the picture window? But if it was the first bullet, then the killer must have been someone Anderson would not let into the house.

McAllister fired into the picture window.

Glass showered inward. There was the scream of an alarm.

McAllister stood rooted. It was a terrible sound, and in these quiet hills it would carry forever! He hadn't expected alarms. There must be a secondary system, continually in operation—Hell with it. McAllister ran into the house, picked up the tablecloth and ran out. Glass particles all over his shoes. Never mind. His shoes and everything else he was wearing were paper, and there was a change of clothing in the briefcase. He'd dump gun and all at the next number he dialed.

The altitude was getting to him. He was panting like a bloodhound when he closed the booth door and dialed. Los Angeles International, then a lakeside resort in New Mexico. The police could hardly search every lake in the country.

Nothing happened.

He dialed again. And again, while the alarm screamed to the hills, *Help! I am being robbed.* When his hand was shaking too badly to dial, he backed out of the glass door and stood looking at the booth.

This hadn't been in any of the outlines.

The booth wouldn't let him out. In all this vastness he was locked in, locked in with the body.

It was two hours before the helicopter from Fresno arrived. Even so, they made good speed. Only a police organization could get a copter in the air that fast. Who else dealt with situations in which one could not simply flick in?

The copter landed in front of the Anderson house, after some trouble picking it out of the wild landscape. Police Lieutenant Richard Donaho climbed out carefully as soon as the dust had stopped swirling. For the benefit of the pilot his face was unnaturally blank. The fear of death had taken him the instant the blades started whirling around, and it was only now leaving him.

With the motor off, the alarm from the house was an intolerable scream. Lieutenant Donaho moved around to the side of the machine, opened a hatch and switched in the portable JumpShift unit.

He stood back as men and equipment began pouring through. Uniformed men moved toward the house, spreading out. Donaho didn't interfere. He wasn't expecting anything startling. It was going to be cold burglary, the burglar vanished quite away.

It was a smallish one-story house in a wild and beautiful setting, halfway up a mountain. The sun was still bright, though it had almost touched the western peaks. The sky was dark blue, almost lavender. Houses were scarce upslope, and far scarcer downslope. There were no roads. No roads at all. This place must have been uninhabited until twenty years ago, when JumpShift Inc. had revolutionized transportation.

The shrill of the alarm stopped.

In the sudden silence a policeman walked briskly from around the side of the house. "Lieutenant!" he called. "It's not burglary. It's murder. There's a dead man on the living room rug."

"All right," said Donaho. He called Homicide.

Captain Hennessey flicked in with the hot summer air of Fresno around him. It puffed out when he opened the door, and he felt the dry chill of the mountains. His ears popped. He stepped out of the belly of the copter, looking for the nearest man. "Donaho! What's happening?"

Donaho nodded at the uniformed man, whose name was Fisher. Fisher said, "It's around in back. Picture window shattered. Man inside, dead, with two holes in his back. That's as far as we've got. Want to come look, sir?"

"In a minute. What was wrong with the displacement booth? Never mind, I see it."

It was obvious even from here. The displacement booth was a standard model, a glass cylinder rounded at the top, with a dial system set in the side. Its curved door was blocked open by a chunk of granite.

"So that's why you needed the copter," said Hennessey, "Hmm." He hadn't expected that.

It was an old trick. Any burglar knew enough to block the displacement booth door before trying to rob a house. If he set off an alarm the police couldn't flick in, and he could generally run next door and use the displacement booth there. But here . . .

"I wonder how he got out?" said Hennessey. "He couldn't set the rock and then use the booth. Maybe he couldn't use the booth anyway. Some alarms lock the transmitter on the booth, so people can still flick in but nobody can flick out."

Donaho shifted impatiently. This was a murder investigation, and he had not yet so much as seen the body.

Hennessey looked down a rocky, wooded slope, darkening with dusk. "Hikers would call this leg-breaker country," he said. "But that's how he did it. There's no other way he could get out. When the booth wouldn't send him anywhere, he blocked the door open and set out for . . . hmm."

The nearest house was half a mile away. It was bigger than Anderson's house, with a pool and a stretch of lawn and a swing and a slide, all clearly visible in miniature from this vantage point.

"For there, I think. He'd rather go down than up. He'd have to circle that stretch of chaparral . . ."

"Captain, do you really think so? I wouldn't try walking through that."

"You'd stay here and wait for the fuzz? It's not *that* bad. You'd make two miles an hour without a backpack. Hell, he might even have planned it this way. I hope he left footprints. We'll want to know if he wore hiking boots." Hennessey scowled. "Not that it'll do us any good. He could have reached the nearest house a good hour ago."

"That doesn't mean he could use the booth. Someone might have seen him."

"Hmm. Right. Or . . . he might have broken an ankle anyway, mightn't he? Donaho, get that copter up and start searching the area. We'll have someone in Fresno question the neighbors. With the alarm blaring like that, they might have been more than usually alert."

Lieutenant Donaho had not greatly enjoyed his first helicopter flight, which had ended twenty minutes ago. Now he was in the air again, and the slender wings were beating round and round over his head, and the ground was an uncomfortable distance below.

"You don't like this much," the pilot said perceptively. He was a stocky man of about forty.

"Not much," Donaho agreed. It would have been nice if he could close his eyes, but he had to keep watching the scenery. There were trees a man could hide in, and a brook a man might have drunk from. He watched for movement; he watched for footprints. The scenery was both too close and too far down, and it wobbled dizzyingly.

"You're too young," said the pilot. "You young ones don't know anything about speed."

Donaho was amused. "I can go anywhere in the world at the speed of light."

"Hell, that isn't speed. Ever been on a motorcycle?"

"No!"

"I was using a chopper when they started putting up the JumpShift booths all over the place. Man, it was wonderful. It was like all the cars just evaporated! It took years, but it didn't seem that way. They left all those wonderful freeways for just us. You know what the most dangerous thing was about riding a chopper? It was cars."

"Yah."

"Same with flying. I don't own a plane. God knows I haven't got the money, but I've got a friend who does. It's a lot more fun now that we've got the airfields to ourselves. No more big planes. No more problem refueling either. We used to worry about running out of gas."

"Uh huh." A thought struck Donaho. "What do you know about off-the-road vehicles?"

"Not that much. They're still made. I can't think of one small enough to fit into a displacement booth, if that's what you're thinking."

"I was. Hennessey thinks the killer might have set off the alarm deliberately. If he did, he might have brought an off-the-road vehicle along. Are you sure he couldn't get one into a booth?"

"No, I'm not." The pilot looked down, considering. "It's too damn steep for a ground-effect vehicle. He'd leave tire tracks."

"What would they look like?"

"Oh, God. You mean it, don't you? Look for two parallel lines, say three to six feet apart. Most tires are corrugated, and you'd see that too."

There was nothing like that in sight.

"Then, I know guys who might try to take a chopper across this. Might break their stupid necks, too. That'd leave a trail like a caterpillar track, but corrugated."

"I can't believe anyone would walk across this. It looked like half a mile of bad stairs back there. And how would he get through those bushes?"

"Crawl. Not that I'd try it myself. But they don't want me for the gas chamber." The pilot laughed. "Can you see the poor bastard, standing in the booth, dialing and dialing—"

Lucas Anderson had been a big man. He had left a big corpse sprawled across a sapphire-blue rug, his arms stretched way out, big hands clutching. Anderson's arms had been dragging a dead weight. One of the holes in his back was high up, just over the spine.

And men moved about him, doing things that would not help him and probably would not catch his killer.

Someone had come here expressly to kill Lucas Anderson. He would have some connection with him, in business or friendship or enmity. He might have left traces of himself, and if he had, these men would find him.

But the alibi machine might have put him anywhere by now. With a valid passport he could be in Algiers or Moscow.

Anderson's bookshelf of his own works showed some science fiction titles. His killer could have been a spaceman—and then he could be in Mars orbit by now, or moving toward Jupiter at lightspeed as a kind of superneutrino.

Yet they were learning things about him.

The cleaning machines had come on as soon as the alarm had been switched off. An alert policeman had got to them before they could do anything about the mess.

There was no glass on the body.

There was no glass under the body either.

"Now, that's not particularly odd," the man in the white coat said to Hennessey. "I mean, the pattern of explosion might have done that. But it means we can't say one way or another."

"He could have been dead when the shot was fired."

"Sure, or the other way around. No glass on him could mean he came running in when he heard all the noise. Just a minute," the man in the white coat said quickly, and he stooped far down to examine Anderson's big shoes with a magnifying glass. "I was wrong. No glass here."

"Hmm. Anderson must have let him in. Then he shot out the window to fox us, and set off the alarm. That wasn't too bright." In a population of three hundred million Americans you could usually find a dozen suspects for any given murder victim. An intelligent killer would simply risk it.

Someday, Hennessey thought when the black mood was on him, someday murder would be an accepted thing. It was that hard to stop. But this one might not have escaped yet ...

"I'd like to get the body to the lab," said the man in the white coat. "Can't do an autopsy here. I want to probe for the bullets. They'd tell us how far away he was shot from, if we can get a gun like it, to do test firing."

"If? Unusual gun?"

The man laughed. "Very. The slug in the wall was a solid-fuel rocket, four nozzles the size of pinholes, angled to spin the thing. Impact like a .45."

"Hmm." Hennessey asked of nobody in particular, "Get any footprints?"

Someone answered. "Yessir, in the grass outside. Paper shoes. Small feet. Definitely not Anderson's."

"Paper shoes." Could he have planned to hike out? Brought a pair of hiking boots to change into? But it began to look like the killer hadn't planned anything so elaborate.

The dining setup would indicate that Anderson hadn't been expecting visitors. If premeditated murder could be called casual, this had been a casual murder, except for the picture window. Police had searched the house and found no sign of theft. Later they could learn what enemies Anderson had made in life. For now ...

For now, the body should be moved to Fresno. "Call the copter back," Hennessey told someone. They'd need the portable JumpShift unit in the side.

When the wind from the copter had died Hennessey stepped forward with the rest, with the team that carried the stretcher. He asked of Donaho, "Any luck?"

"None," said Lieutenant Donaho. He climbed out, stood a moment to feel solid ground beneath his feet. "No footprints, no tracks, nobody hiding where we could see him. There's a lot of woods where he could be hiding, though. Look, it's after sunset, Captain. Get us an infrared scanner and we'll go up again when it gets dark."

"Good." More time for the killer to move—but there were only half a dozen houses he could try for, Hennessey thought. He could get permission from the owners to turn off their booths for awhile. Maybe.

"But I don't believe it," Donaho was saying. "Nobody could travel a mile through that. And the word from Fresno is that the only unoccupied house is two miles off to the side!"

"Never a boy scout, were you?"

"No. Why?"

"We used to hike these hills with thirty pounds of backpack. Still . . . hmm." He seemed to be studying Donaho's face. "Is Anderson's booth back in operation?"

"Yes. You were right, Captain. It was hooked to the alarm."

"Then we can send the copter home and use that. Listen, Donaho, I may have been going at this wrong. Let me ask you something . . ."

Most of the police were gone by ten. The body was gone. There was fingerprint powder on every polished surface, and glass all over the living room.

Hennessey and Donaho and the uniformed man named Fisher sat at the dining table, drinking coffee made in the Anderson kitchen.

"Guess I'll be going home," Donaho said presently. He made no reference to what they had planned.

They watched through the window as Lieutenant Donaho, brilliantly lighted, vanished within the glass booth.

After that they drank coffee, and talked, and watched. The stars were very bright.

It was almost midnight before anything happened. Then, a rustling sound—and something burst into view from upslope, a shadowy figure in full flight. It was in the displacement booth before Hennessey and Fisher had even reached the front door.

The booth light showed every detail of a lean dark man in a rumpled paper business suit, one hand holding a briefcase, the other dialing frantically. Dialing again, while one eye in a shyly averted face watched two armed men strolling up to the booth.

"No use," Hennessey called pleasantly. "Lieutenant Donaho had it cut off as soon as he flicked out."

The man released a ragged sigh.

"We want the gun."

The man considered. Then he handed out the briefcase. The gun was in there. The man came out after it. He had a beaten look.

"Where were you hiding?" Hennessey asked.

"Up there in the bushes, where I could see you. I knew you'd turn the booth back on sooner or later."

"Why didn't you just walk down to the nearest house?"

The lean man looked at him curiously. Then he looked down across a black slope, to where a spark of light showed one window still glowing in a distant house. "Oh my God. I never thought of that."

The Last Days
of the
Permanent Floating Riot Club

In its heyday the Club had numbered around ninety, and it was the most exclusive club in the world. Now a third of its members had quit, and a third were in prison or awaiting trial, and the remaining thirty-odd active members had lost a crucial something: confidence, enthusiasm, esprit de corps, call it what you will.

"We always knew it was coming," said Benny Sherman. He was a thick-set man, short and broad, made mostly of black hair and muscle. He waved a big, stubby-fingered hand at the south wall of the main room, where a commentator was spreading news of the outside world across a wall-sized screen. "It was all over that screen, for years. Central Riot Control in Nebraska. Pictures of the building going up. They told us just how it was gonna work. They gave us a completion date. Twenty of us quit that same day."

Nobody said anything. The voice of the commentator came through at low volume, speaking of the rumor that the Soviets had developed a self-teleporting spy cloak. The teevee screen was never off in the Permanent Floating Riot Club.

"That spy cloak," James Get-It-All (Goethals) said wistfully. "That'd be nice to have when a flash crowd goes sour. I wonder what are the chances of stealing one."

"Sure," said Willie Lordon. He was a featherweight, pinch-faced man, all birdlike bones and acid sarcasm. "Cops coming at you from all directions. What do you do? You roll yourself up in your spy cloak, and as soon as it forms a closed surface it's a displacement booth. Where are you now?" He paused for effect. "In a top secret headquarters in the Kremlin! You idiot."

"Better that than Central Riot Control."

Willie snorted.

"I've been there," said Benny Sherman. "Inside it's like a Rose Bowl without seats. Receiver booths all around the lip of the bowl. You try to flick out of a place where the riot control is on, and you wind up dropping out the bottom of the booth. You slide all the way down to the bottom of the bowl, and you wait there with everyone else till the cops get around to you. I got out by the skin of my teeth."

"By throwing away your take," said Willie Lordon. Clearly the idea disgusted him.

"It hurt, too. I had a diamond the size of an almond, if it was real, and a half dozen good watches . . . and there wasn't any way to tell we'd gone on riot control. I just had to guess the flash crowd had gone on long enough."

"You're a genius," said Willie.

"I'm losing my nerve," Benny said mournfully. "Six times this past year we've flicked into flash crowds, and three times I threw away everything I had because it looked like the cops had time to put us under riot control. Once I was right. Twice I was wrong. That's just not good enough." He braced himself. "I think I'll quit." There, he'd said it.

"Shh," said Lou Garcia, waving them to silence. He turned the volume control louder. The teevee newscaster was saying, ". . . flash crowd in downtown Topeka seems to have developed due to a heavily advertised sale at Bloomingdale's . . ."

"Shh, Hell. I quit!" Benny bellowed over the racket. "We made a lot of money the last ten years. I want to stay outside to enjoy it!"

Most of the members were on their feet, eyes on the screen. A flash crowd meant business. James Get-It-All was at the computer terminal getting the numbers of displacement booths in the affected area. An endless strip of paper ran from the slot: thirty-odd copies of the list.

Lou Garcia favored Benny with a sardonic look. "You're giving up your share of the treasury?"

That was a low blow. Benny stood a moment, considering. Then, "You can have it," he said, and walked out.

He turned for one last look at the Club before going on. It seemed likely that he would never see it again.

The Club was a three story brick building of prestressed concrete made to look like old brick. The brick/concrete was chipped in spots and dark with age: one among several blocks of older buildings. The luxuries were inside: luxuries bought with Club dues.

Now other members were filing out the entrance and dividing there, heading for street-corner displacement booths half a block away in each direction. Willie Lordon was flexing his fingers as he walked. He carried a small electric knife that would slice out the bottom of a citizen's lock pocket, without alerting him if there was sufficient noise and jostling to distract him. James Get-It-All jogged along with the tense, serious look of a player who knows that his team depends on him. Lou Garcia stood at the entrance, grinning broadly as he watched them go.

They filed into glass cylinders with rounded tops, dialed and disappeared, one by one.

Benny watched them wistfully. He had helped to found the Club, and they didn't even know he was gone.

He remembered a September night ten years ago, the night Orrie Black had proposed the idea. He and Orrie and Lou Garcia and some others who had gotten their start when the booths were new. In those days you could get the booth number of a house and just flick into the living room or entrance hall. You could make a strike just by dialing at random until you hit. But the citizens had wised up and started putting their booths outside, and now half a dozen ex-burglars had gathered at a topless beer and pool place.

"Think it out," Orrie Black had said. "Any time something interesting happens, anywhere in the country, some newstaper is going to report it. If it's interesting enough, people are going to flick in to see it, *from all over the country*. Now just think about that. With these long distance booths you can get from anywhere to anywhere else just by dialing three numbers.

"If the crowd gets big enough a lot more people flick in just to see a flash crowd, plus more newstapers, plus any kind of agitator looking to shove his sign in front of a camera, plus looters and pickpockets and cops. Before anyone knows it you've got a riot going, with everyone breaking windows and grabbing what's in them. So why shouldn't we be the ones breaking windows?

"The key, the crucial thing, is for there to be enough of us to help each other out. We should all be flicking in at once . . ." And they'd tried it out in the Third Watts Riot, which had lasted a full day and a half.

These days you were lucky if a flash crowd lasted two hours. And Orrie Black was in prison, and the others had gone their ways—all but Benny and Lou Garcia.

The Club dues. Not everyone had liked that idea, Benny included. Half your take! But it had paid off, and not just for the Clubhouse. The treasury had paid defense lawyers and hospital fees. Flicking into a riot was dangerous, even for a pro.

There must be a lot of it left in Lou Garcia's care. Quitting had cost Benny his share of that.

Still—he shuddered, remembering the last one. Despite previous experience, he hadn't expected it to grow so big so fast.

Something trivial had started it, as usual. A line of people waiting for tickets to a top game show had gotten out of hand. Too many people, not enough seats, somebody getting pushy, and Wham! A pocket riot, until it hit the news, and then a few hundred more flicked in to see the damn fools fighting.

Benny had flicked into the middle of it, already looking around for the stores—and the cops. The cops had learned something in past years. It wasn't that there were so many of them: it was their deployment. They tended to guard the most valuable store windows. Benny had spotted a furrier's, a small jewelry display, a home appliance store—all guarded by cops. He had seen clerks moving within the furrier's window, trying to get the goods out of harm's way.

He had pushed his way out into the swarm: newstapers with gyrostabilized cameras, a scattering of hand-lettered signs held high, and a hell of a lot of people caught up in it somehow, unable to flick out because the displacement booths filled with incoming passengers before anyone could get in. A lot of incoming passengers had been Club members. A normal enough crowd, but so thick!

The crowd had surged suddenly, downing the cops in front of Van Cleef and Arpel's. Benny had seen the small, wiry man who smashed the window, and scooped, and began pushing his way frantically toward the nearest booth. Toward Benny. And he was not a Club member.

As he passed Benny, Benny had hit him in the stomach and rifled the man's pockets while he was still doubled over. He'd had to fight to keep his feet, but the crowding was such that nobody had noticed what he was doing.

The crowd had surrounded the booth before he turned around. Benny had glimpsed a pair of cops holding back the crowd, letting them into the booth one at a time.

The next nearest booth was a block away, through an incredible sea of feet and elbows. His squat, massive body had been an advantage as he plowed through it. Long before he reached the booth Benny had noticed that nobody was flicking in any more. He had dropped the ring and watches then. Regretfully.

He remembered the sickening moment just after dialing, when the hinged bottom dropped out of the booth and he was sliding downhill. Others were sliding after him, all around the rim of the bowl, and there were hundreds at the bottom picking themselves up, some looking relieved, some furious. The cops had been on a raised, railed platform at the center of the bowl. A loudspeaker had been telling the crowd what they already knew: that they were at United States Central Riot Control, that they would be processed as fast as possible and released.

The police had searched him, photographed him, and sent off the photos for comparison with records of previous riots. His face was on record: he had been in other flash crowds.

They had held him. They had held quite a lot of people, many of whom weren't even Club members.

"Just a coincidence," he had told the police. "It's funny how many flash crowds I run into. Never been hurt in one, though. I guess I'm lucky."

They couldn't prove different. They'd had to let him go.

But they knew. Benny hunched his big shoulders, remembering the contempt in their eyes. They knew. And his face and fingerprints had been caught in one more flash crowd. They'd get him if he kept it up.

It was time to quit.

What about the treasury? When most of the members had quit or been caught and sent up, would it be divided by the last few? Lou Garcia must think so. That was why he had gone with the others. That was why he was grinning.

Benny couldn't bring himself to like the idea. He had collected his share of the treasury. But what could he do? If he stayed in the Club but avoided the flash crowds, the others would get tired of collecting his share of the dues for him. They'd beat him up and kick him out.

It had happened before. Club activities depended on there being enough members in a flash crowd to help each other. Goldbricks were not tolerated.

He stood in a corner booth, coin in hand, wondering where he wanted to go. Where to go, when a career has ended? What difference does it make? The flash crowd at Bloomingdale's was actually in walking distance, and he was tempted to go watch. The police barricades must be up by now. He could look across them, watch the Club in action.

No flash crowd had ever happened this close to the Club. A good thing it hadn't happened nearer . . .

The idea came to him that suddenly.

For Jerryberry Jansen, home was two rooms knocked together in what had once been a motel on the Pacific Coast Highway. The rooms sold as apartments now. They were cheap, and there was a swimming pool and access to the ocean. The concrete walk between the two rows of doors still had fading white diagonal lines on it.

Five o'clock found Jerryberry flopped bonelessly across the double bed.

For six years Jerryberry had been one of CBA's wandering newstapers, whose profession it is to flick about Los Angeles without leaving the booths, carrying a hand-held camera in hope of finding something interesting to report. He had developed legs like tree trunks. These days he went out on assignment: a step up, but it still involved legwork. Some day, he thought as he put his feet up on the pillows, JumpShift Inc. would start putting seats in the booths. But first they'd have to figure out how to flick the passenger out without flicking the seats out too.

The phone rang.

First he cursed. Then he heaved himself upright and put on a smile to answer. The smile sagged when the screen remained blank. A voice said, "Barry Jerome Jansen?"

"Speaking."

"The newstaper?"

"Right again. Who's this?" Jerryberry wondered if it was a crank call. The voice belonged on a bad actor playing the role of a tough.

"It doesn't matter who I am. How would you like the address of the Permanent Floating Riot Club?"

Jerryberry checked his first response, which would have been, "I'd love it." He said, "There isn't any such thing." That response was justified too. Nobody had ever proved the existence of a Permanent Floating Riot Gang. Every flash crowd attracted a certain proportion of looters. So what?

But he flipped a switch to record the call. The voice had said *Club*, not *Gang*.

"There is too," it said impatiently. "It's at 225 East Lindon, Topeka."

"You're not trying to sell it?"

"I'm *giving* it to you, baby. Did you get it? 225 East Lindon Drive, Topeka, Kansas." The caller hung up.

Jerryberry flopped back on the bed. He was tired. It could be a gag. Topeka, Kansas. Who would be telling Jerryberry Jansen about it? Jerryberry's beat was Los Angeles.

Oh, well. He heaved himself upright and called the police.

The Topeka police were spending all their time answering the phone. "We know," said Detective Sergeant Hirohito. "That's the same address he gave everyone. Thank you for calling; we're already on it." He hung up. "Another one. Los Angeles. He must have called every newscaster in the country."

"God, I hope not. They won't all keep their mouths shut. We've got to have time to check this out."

Hirohito drummed his fingers on the desk. "There's only one way to get it. We'll have to put the whole area under riot control."

"What? No. If it's a false alarm, we could get sued for obstructing business! There are a lot of mail order houses in the area, not to mention a messenger service—"

"Calm down, Jack. Now we both know this is going to hit the news sooner or later, probably about now. What's going to happen then?"

Jack Shorter grinned. "Sure. Flash crowd!"

"It'll be the first time we ever put the riot control on before the riot started. The newstapers'll probably call it the Riot Club Riot."

Most of the news programs reported the incident along with a bulletin from the Topeka Police Force. *We have not yet had time to erect barricades, and the suspects could be armed. We strongly advise citizens to stay out of the affected area . . .*

"They always say that," CBA's commentator, Wash Evans, told his audience. "But you never pay any attention. This time they mean it. There's no telling what kind of weaponry a Looters' Club might have picked up in the last ten years. We know they've raided a few sporting goods stores in there, and there have been a few shoot-outs. Do not go to see the riot. You get a better view on teevee."

Nobody paid any attention.

Central Riot Control. The theory was simple enough. You divided all of the municipal areas in the United States into areas of approximately four blocks by four blocks. Outside the cities the areas were far bigger, the flash crowds far less likely.

When a flash crowd gathered, there were switches at the police stations that would affect all of the displacement booths in one or more riot control areas. With riot control going, the booths in the area would not admit incoming passengers except from the police stations. They would send only to the huge Central Riot Control Building in Nebraska.

The Permanent Floating Riot Club kept maps of most of the riot control areas in the country. There were tens of thousands of them, and they were stored in an expensive computer on the third floor of the Club.

In simple curiosity, Benny had once looked up the area the Club itself was situated in. He had been amused to find that Lou Garcia—who lived three blocks away—was in the same riot control area. Lou may have done that deliberately. If the Club was ever put under Riot Control, he could simply stroll home.

He was going to regret that bit of cleverness.

Benny had not called every newscaster in the country. It would have taken too long. He had called about twenty of the most famous. Now he hung up and strolled out into the street.

This area hadn't changed much over the past decade. In fact, that was true of most municipal areas. The new buildings were all going up in rural and desert areas, where men could work and live with more elbow room and prettier scenery than their city cousins, without sacrificing anything in the way of mobility. Here in the civic center the buildings just sat there growing older: brick and concrete darkening with smog, small buildings growing grimy. The people were generally older too. Benny had once noticed that you could tell a citizen's age further away than you could tell his sex, by the tenacity with which he hugged the sidewalk instead of strolling down the center of the street, or by whether he looked both ways for phantom cars before crossing.

As he crossed an intersection Benny glimpsed the Club building three blocks down. Nothing happening there. And there were no barricades yet. But there were people leaving nearby booths, flicking in at a good rate, it seemed, and they all walked like young men.

He entered Lou Garcia's apartment building and rang Garcia's bell in the lobby.

It seemed pretty well foolproof at this stage. If Garcia wasn't home, then he was either at the Club or elsewhere. If he was at the Club, they'd hold him. If he was somewhere else, he wouldn't be able to flick in. The cops must have put this area under riot control by now. In either case, Benny would have time to search his apartment. He had been in Garcia's apartment many times. There was a hall closet that Garcia always kept locked . . .

"Yah?" The intercom.

"Benny. Can I see you?"

Hesitation. Then, "Sure. Come on up." The main door buzzed open.

Well, he was home, and it was going to be a little sticky. It would still work out. Lou couldn't flick out now even if he got past Benny.

Benny had a gun in his hand as the elevator opened. There was nobody in the hall. Benny walked down to Lou's door and rapped.

"Just a minute," Lou Garcia sang out from inside.

Benny's mind flashed ahead. Suppose the money wasn't in Lou's apartment? Well, that would be that. But Garcia wouldn't keep the money in a bank. He wouldn't dare. And there was that permanently locked closet. And he'd always had the money available when needed. And . . . well, it was a gamble.

He mumbled words under his breath, rehearsing what amounted to a speech. "Someone blew the whistle on us," he would say. "Someone gave the cops the Club address. I'll tell them it was you. Hell, they'll probably figure that out for themselves. You're the only one who had anything to gain. I'll tell them you were running off with the Club treasury. You can't flick out," he would say. "Half the Club must have been at the Bloomingdale's flash crowd when the riot control came on. They'll come trickling in looking for you. But if you give me half the treasury—" Better settle for a third. Damn, if Lou had been out he could have searched the apartment and had it all. He could still do that if he were to shoot Garcia. But he'd known Lou too long for that.

"A third of the treasury, and we just wait till riot control goes off. Then we flick out in separate directions. Dial at random, settle wherever we land, live on the money the rest of our lives. Who could find us?"

It was taking Garcia a long time.

Benny kicked at the door. "Open up, Lou!" He kicked harder, and the door flew wide, and Benny ducked to the side just in case. No bullets. He went through fast, but nothing happened. Lou Garcia wasn't in sight.

He wasn't in either bathroom. He wasn't in the kitchen or on the balcony. Benny tried the closets last. The one that had always been locked opened easily, and there was nothing inside at all.

So. Lou had gotten out. (How? There was only the one door.) Which left Benny to search the apartment in peace. Unless Lou had *taken* the treasury . . .

Benny peered over the balcony. Lou could have reached the street by now . . . but he wasn't in sight. He might have been hidden by the milling crowd below. The flash crowd was developing nicely. As Benny had expected, they had come flicking in from all over, arriving outside the affected riot control areas and strolling in to see the excitement.

If the cops found Benny now, he'd claim he was one of them. He'd flicked in to watch the arrests. But the same went for Lou, unless Lou was carrying the treasury, in which case he might have some explaining to do.

So. It might still be here. Benny started his search . . . and stopped, bewildered. There were other peculiarities. Things missing. Like: the big reading chair was still here, and the heavy coffee table. But the little fold-up chairs and the water bed were gone, and the tall reading lamp . . . Benny looked around, trying to puzzle it out. It was as if Lou were halfway through moving . . . as if he had been taking only those things that would fit into a displacement booth.

Benny saw it then, and he ran for the closet.

The closet that had always been locked. A closet like a cylinder with a rounded top, the curve continued on the inside of the door. And nothing at all inside.

It was a displacement booth.

Benny started to laugh. Lou had thought of it first. He was

planning to disappear with the treasury; but he didn't know the area was under riot control.

Of course Benny could search the apartment anyway. But Lou wouldn't have left the money behind, not with Benny standing on the other side of the door.

Benny set the gun down on the remaining table. Where he expected to be going, it was a danger to him. He stepped into the closet and closed the door. There was no dial in here. It must have a preset destination.

Light flashed in his eyes, and the floor opened beneath him. Benny had been through this before. He took the fall like an amusement park ride, and stood up when it was over.

Central Riot Control was crowded today. Citizens milled about the floor of the great bowl, making angry noises, hampered by the attempts of Club members to look inconspicuous among them. There were too many Club members and not enough citizens. It took Benny only a moment to find Lou.

Lou was in a clump of people to one side of the big central platform where the cops waited. He was trying to hold onto a sizeable metal attaché case, and four members of the Permanent Floating Riot Club were trying to take it from him. The cops on the platform watched with interest.

Benny sighed. It grieved him to see ten years of history ending. But he still had fifty percent of ten years' earnings . . . and it had been worth a try.

A Kind of Murder

"You are constantly coming to my home!" he shouted. "You never think of calling first. Whatever I'm doing, suddenly you're there. And where the hell do you keep getting keys to my door?"

Alicia didn't answer. Her face, which in recent years had taken on a faint resemblance to a bulldog's, was set in infinite patience as she relaxed at the other end of the couch. She had been through this before, and she waited for Jeff to get it over with.

He saw this, and the dinner he had not quite finished settled like lead in his belly. "There's not a club I belong to that you aren't a member too. Whoever I'm with, you finagle me into introducing you. If it's a man, you try to make him, and if he isn't having any you get nasty. If it's a woman, there you are like the ghost at the feast. The discarded woman. It's a drag," he said. He wanted a more powerful word, but he couldn't think of one that wouldn't sound overdramatic, silly.

She said, "We've been divorced six years. What do you care who I sleep with?"

"I don't like looking like your pimp!"

She laughed.

The acid was rising in his throat. "Listen," he said, "why

53

don't you give up one of the clubs? We, we belong to four. Give one up. Any of them." Give me a place of refuge, he prayed.

"They're my clubs too," she said with composure. "You change clubs."

He'd joined the Lucifer Club four years ago, for just that reason. She'd joined too. And now the words clogged in his throat, so that he gaped like a fish.

There were no words left. He hit her.

He'd never done that before. It was a full-arm swing, but awkward because they were trying to face each other on the couch. She rode with the slap, then sat facing him, waiting.

It was as if he could read her mind. *We've been through this before, and it never changes anything. But it's your tantrum.* He remembered later that she'd said that to him once, those same words, and she'd looked just like that: patient, implacable.

The call reached Homicide at 8:36 P.M., July 20, 2019. The caller was a round-faced man with straight black hair and a stutter. "My ex-wife," he told the desk man. "She's dead. I just got home and f-found her like this. S-someone seems to have hit her with a c-c-cigarette box."

Hennessey (Officer-2) had just come on for the night shift. He took over. "You just got home? You called immediately?"

"That's right. C-c-could you come right away?"

"We'll be there in ten seconds. Have you touched anything?"

"No. Not her, and not the box."

"Have you called a hospital?"

His voice rose. "No. She's *dead.*"

Hennessey took down his name—Walters—and booth number and hung up. "Linc, Fisher, come with me. Torrie, will you call the City Hospital and have them send a 'copter?" If Walters hadn't touched her he could hardly be sure she was dead.

They went through the displacement booth one at a time, dialing and vanishing. For Hennessey it was as if the Homicide room vanished as he dialed the last digit, and he was looking into a porch light.

Jeffrey Walters was waiting in the house. He was medium sized, a bit overweight, his light brown hair going thin on top. His paper business suit was wrinkled. He wore an anxious, fearful look—which figured, either way, Hennessey thought.

And he'd been right. Alicia Walters was dead. From her attitude she had been sitting sideways on the couch when something crashed into her head, and she had sprawled forward. A green cigarette box was sitting on the glass coffee table. It was bloody along one edge, and the blood had marked the glass.

The small, bloody, beautifully marked green malachite box could have done it. It would have been held in the right hand, swung full-armed. One of the detectives used chalk to mark its position on the table, then nudged it into a plastic bag and tied the neck.

Walters had sagged into a reading chair as if worn out. Hennessey approached him. "You said she was your ex-wife?"

"That's right. She didn't give up using her married name."

"What was she doing here, then?"

"I don't know. We had a fight earlier this evening. I finally threw her out and went back to the Sirius Club. I was half afraid she'd just follow me back, but she didn't. I guess she let herself back in and waited for me here."

"She had a key?"

Walters' laugh was feeble. "She always had a key. I've had the lock changed twice. It didn't work. I'd come home and find her here. 'I just wanted to talk,' she'd say." He stopped abruptly.

"That doesn't explain why she'd let someone else in."

"No. She must have, though, mustn't she? I don't know why she did that."

The ambulance helicopter landed in the street outside. Two men entered with a stretcher. They shifted Alicia Walters' dead body to the stretcher, leaving a chalk outline Fisher had drawn earlier.

Walters watched through the picture window as they walked the stretcher into the portable JumpShift unit in the side of the 'copter. They closed the hatch, tapped buttons in a learned rhythm on a phone dial set in the hatch. When they opened the hatch to check, it was empty. They closed it again and boarded the 'copter.

Walters said, "You'll do an autopsy immediately, won't you?"

"Of course. Why do you ask?"

"Well . . . it's possible I might have an alibi for the time of the murder."

Hennessey laughed before he could stop himself. Walters looked puzzled and affronted.

Hennessey didn't explain. But later, as he was leaving the station house for home and bed, he snorted. "Alibi," he said. "Idiot."

The displacement booths had come suddenly. One year, a science fiction writer's daydream. The next, A.D. 1992, an experimental reality. Teleportation. Instantaneous travel. Another year and they were being used for cargo transport. Two more, and the passenger displacement booths were springing up everywhere in the world.

By luck and the laws of physics, the world had had time to adjust. Teleportation obeyed the Laws of Conservation of Energy and Conservation of Momentum. Teleporting uphill took an energy input to match the gain in potential energy. A cargo would lose potential energy going downhill. And it was over a decade before JumpShift Inc. learned how to compensate for that effect. Teleportation over great distances was even more heavily restricted by the Earth's rotation.

Let a passenger flick too far west, and the difference between his momentum and the Earth's would smack him down against the floor of the booth. Too far east, and he would be flung against the ceiling. Too far north or south, and the Earth would be rotating faster or slower; he would flick in moving sideways, unless he had crossed the equator.

But cargo and passengers could be displaced between points of equal longitude and opposite latitude. Smuggling had become impossible to stop. There was a point in the South Pacific to correspond to any point in the United States, most of Canada, and parts of Mexico.

Smuggling via the displacement booths was a new crime. The Permanent Floating Riot Gangs were another. The booths would allow a crowd to gather with amazing rapidity.

Practically any news broadcast could start a flash crowd. And with the crowds the pickpockets and looters came flicking in.

When the booths were new, many householders had taken to putting their booths in living rooms or entrance halls. That had stopped fast, after an astounding rash of burglaries. These days only police stations and hospitals kept their booths indoors.

For twenty years the booths had not been feasible over distances greater than ten miles. If the short distance booths had changed the nature of crime, what of the long distance booths? They had been in existence only four years. Most were at what had been airports, being run by what had been airline companies. Dial three numbers and you could be anywhere on Earth.

Flash crowds were bigger and more frequent.

The alibi was as dead as the automobile.

Smuggling was cheaper. The expensive, illegal transmission booths in the South Pacific were no longer needed. Cutthroat competition had dropped the price of smack to something the Mafia wouldn't touch.

And murder was easier; but that was only part of the problem. There was a new *kind* of murder going around.

Hank Lovejoy was a tall, lanky man with a lantern jaw and a ready smile. The police had found him at his office—real estate—and he had agreed to come immediately.

"There were four of us at the Sirius Club before Alicia showed up," he said. "Me, and George Larimer, and Jeff Walters, and Jennifer—wait a minute—Lewis. Jennifer was over at the bar, and we'd like asked her to join us for dinner. You know how it is in a continuity club: you can talk to anyone. We'd have picked up another girl sooner or later."

Hennessey said, "Not two?"

"Oh, George is a monogamist. His wife is eight months pregnant, and she didn't want to come, but George just doesn't. He's not fey or anything, he just doesn't. But Jeff and I were both sort of trying to get Jennifer's attention. She was loose, and it looked likely she'd go home with one or the other of us. Then Alicia came in."

"What time was that?"

"Oh, about six fifteen. We were already eating. She came up to the table, and we all kind of waited for Jeff to introduce her and ask her to sit down, she being his ex-wife, after all." Lovejoy laughed. "George doesn't really understand about Jeff and Alicia. Me, I thought it was funny."

"What do you mean?"

"Well, they've been divorced about six years, but it seems he just can't get away from her. Couldn't, I mean," he said, remembering. Remembering that good old Jeff *had* gotten away from her, because someone had smashed her skull.

Hennessey was afraid Lovejoy would clam up. He played stupid. "I don't get it. A divorce is a divorce, isn't it?"

"Not when it's a quote friendly divorce unquote. Jeff's a damn fool. I don't think he gave up sleeping with her, not right after the divorce. He wouldn't live with her, but every so often she'd, well, she'd seduce him, I guess you'd say. He wasn't used to being alone, and I guess he got lonely. Eventually he must have given that up, but he still couldn't get her out of his hair.

"See, they belonged to all the same clubs and they knew all the same people, and as a matter of fact they were both in routing and distribution software; that was how they met. So if she came on the scene while he was trying to do something else, there she was, and he had to introduce her. She probably knew the people he was dealing with, if it was business. A lot of business gets done at the continuity clubs. And she wouldn't go away. I thought it was funny. It worked out fine for me, last night."

"How?"

"Well, after twenty minutes or so it got through to us that Alicia wasn't going to go away. I mean, we were eating dinner, and she wasn't, but she wanted to talk. When she said something about waiting and joining us for dessert, Jeff stood up and suggested they go somewhere and talk. She didn't look too pleased, but she went."

"What do you suppose he wanted to talk about?"

Lovejoy laughed. "Do I read minds without permission? He wanted to tell her to bug off, of course! But he was gone half an hour, and by the time he came back Jennifer and I had sort of reached a decision. And George had this benign look he gets, like *Bless you my children*. He doesn't play around himself, but maybe he likes to think about other couples getting together. Maybe he's right; maybe it brightens up the marriage bed."

"Jeff came back alone?"

"That he did. He was nervous, jumpy. Friendly enough; I mean, he didn't get obnoxious when he saw how it was with me and Jennifer. But he was sweating, and I don't blame him."

"What time was this?"

"Seven twenty."

"Dead on?"

"Yah."

"Why would you remember a thing like that?"

"Well, when Jeff came back he wanted to know how long he'd been gone. So I looked at my watch. Anyway, we stayed another fifteen minutes and then Jennifer and I took off."

Hennessey asked, "Just how bad were things between Jeff and Alicia?"

"Oh, they didn't fight or anything. It was just ... funny. For one thing, she's kind of let herself go since the divorce. She used to be pretty. Now she's gone to seed. Not many men chase her these days, so she has to do the chasing. Some men like that."

"Do you?"

"Not particularly ... I've spent some nights with her, if that's what you're asking. I just like variety. I'm not a heartbreaker, man; I run with girls who like variety too."

"Did Alicia?"

"I think so. The trouble was, she slept with a lot of guys Jeff introduced her to. He didn't like that. It made him look bad. And once she played nasty to a guy who turned her down, and it ruined a business deal."

"But they didn't fight."

"No. Jeff wasn't the type. Maybe that's why they got divorced. She was just someone he couldn't avoid. We all know people like that."

"After he came back without Alicia, did he leave the table at any time?"

"I don't think so. No. He just sat there, making small talk. Badly."

George Larimer was a writer of articles, one of the few who made good money at it. He lived in Arizona. No, he didn't mind a quick trip to the police station, he said, emphasizing the *quick*. Just let him finish this paragraph—and he breezed in five minutes later.

"Sorry about that. I just couldn't get the damn wording right. This one's for *Viewer's Digest*, and I have to explain drop ship technology for morons without talking down to them or the minimal viewer won't buy it. What's the problem?"

Hennessey told him.

His face took on an expression Hennessey recognized: like he ought to be feeling something, and he was trying, honest. "I just met her that night," he said. "Dead. Well."

He remembered that evening well enough. "Sure, Jeff Walters came back about the time we were finishing coffee. We had brandy with the coffee, and then Hank and, uh, Jennifer left. Jeff and I sat and played dominos over Scotch and sodas. You can do that at the Sirius, you know. They keep game boxes there, and they'll move up side tables at your elbows so you can have drinks or lunch."

"How did you do?"

"I beat him. Something was bothering him; he wasn't playing very well. I thought he wanted to talk, but he wouldn't talk about whatever was bugging him."

"His ex-wife?"

"Maybe. Maybe not. I'd only just met her, and she seemed nice enough. And she seemed to like Jeff."

"Yah. Now, Jeff left with Alicia. How long were they gone?"

"Half an hour, I guess. And he came back without her."

"What time?"

"Quarter past seven or thereabouts. Ask Hank. I don't wear a watch." He said this with a certain pride. A writer doesn't need a watch; he sets his own hours. "As I said, we had dessert and coffee and then played dominos for an hour, maybe a little less. Then I had to go home to see how my wife was getting along."

"While you were having dessert and coffee and playing dominos, did Jeff Walters leave the table at any time?"

"Well, we switched tables to set up the game." Larimer shut his eyes to think. He opened them. "No, he didn't go to the bathroom or anything."

"Did you?"

"No. We were together the whole time, if that's what you want to know."

Hennessey went out for lunch after Larimer left. Returning, he stepped out of the Homicide Room booth just ahead of Officer-1 Fisher, who had spent the morning at Alicia Walters' place.

Alicia had lived in the mountains, within shouting distance of Lake Arrowhead. Property in that area was far cheaper than property around the Lake itself. The high rent district in the mountains is near streams and lakes. Her own water supply had come from a storage tank kept filled by a small JumpShift unit.

Fisher was hot and sweaty and breathing hard, as if he had been working. He dropped into a chair and wiped his forehead and neck. "There wasn't much point in going," he said. "We found what was left of a bacon and tomato sandwich sitting on a placemat. Probably her last meal. She wasn't much of a housekeeper. Probably wasn't making much money, either."

"How so?"

"All her gadgetry is old enough to be going to pieces. Her Dustmaster skips corners and knocks things off tables. Her chairs and couches are all blow-ups, inflated plastic. Cheap, but they have to be replaced every so often, and she didn't. Her displacement booth must be ten years old. She should have replaced it, living in the mountains."

"No roads in that area?"

"Not near her house, anyway. In remote areas like that they move the booths in by helicopter, then bring the components for the house out through the booth. If her booth broke down she'd have had to hike out, unless she could find a neighbor home, and her neighbors aren't close. I like that area," Fisher said suddenly. "There's elbow room."

"She should have made good money. She was in routing and distribution software." Hennessey pondered. "Maybe she spent all her time following her ex-husband around."

The autopsy report was waiting on his desk. He read through it.

Alicia Walters had indeed been killed by a single blow to the side of the head, almost certainly by the malachite box. Its hard corner had crushed her skull around the temple. Malachite is a semiprecious stone, hard enough that no part of it had broken off in the wound; but there was blood and traces of bone and brain tissue on the box itself.

There was also a bruise on her cheek. *Have to ask Walters about that,* he thought.

She had died about 8:00 P.M., given the state of her body, including body temperature. Stomach contents indicated that she had eaten about 5:30 P.M.: a bacon and tomato sandwich.

Hennessey shook his head. "I was right. He's still thinking in terms of alibis."

Fisher heard. "Walters?"

"Sure, Walters. Look: he came back to the Sirius Club at seven twenty, and he called attention to the time. He stayed until around eight thirty, to hear Larimer tell it, and he was always in someone's company. Then he went home, found the body and called us. The woman was killed around eight, which is right in the middle of his alibi time. Give or take fifteen minutes for the lab's margin of error, and it's still an alibi."

"Then it clears him."

Hennessey laughed. "Suppose he did go to the bathroom. Do you think anyone would remember it? Nobody in the world has had an alibi for anything since the JumpShift booths took over. You can be at a party in New York and kill a man in the California Sierras in the time it would take to go out for cigarettes. You can't use displacement booths for an alibi."

"You could be jumping to conclusions," Fisher pointed out. "So he's not a cop. So he reads detective stories. So someone murdered his wife in his own living room. *Naturally* he wants to know if he's got an alibi."

Hennessey shook his head.

"She didn't bleed a lot," said Fisher. "Maybe enough, maybe not. Maybe she was moved."

"I noticed that too."

"Someone who knew she had a key to Walters' house killed her and dumped her there. He would have hit her with the cigarette box in the spot where he'd already hit her with something else."

Hennessey shook his head again. "It's not just Walters. It's a *kind* of murder. We get more and more of these lately. People kill each other because they can't move away from each other. With the long distance booths everyone in the country lives next door to everyone else. You live a block away from your ex-wife, your mother-in-law, the girl you're trying to drop, the guy who lost money in your business deal and blames you. Any secretary lives next door to her boss, and if he needs something done in a hurry she's right there. God help the doctor if his patients get his home number. I'm not just pulling these out of the air. I can name you an assault rap for every one of these situations."

"Most people get used to it," said Fisher. "My mother used to flick in to visit me at work, remember?"

Hennessey grinned. He did. Fortunately, she'd given it up. "It was worse for Walters," he said.

"It didn't really sound that bad. Lovejoy said it was a friendly divorce. So he was always running into her. So what?"

"She took away his clubs."

Fisher snorted. But Fisher was young. He had grown up with the short-distance booths.

For twenty years passenger teleportation had been restricted to short hops. People had had time to get used to the booths. And in those twenty years the continuity clubs had come into existence.

The continuity club was a guard against future shock. Its location was ... ubiquitous: hundreds of buildings in hundreds of cities, each building just like all the others, inside and out. Wherever a member moved in this traveling society, the club would be there. Today even some of the customers would be the same: everyone used the long distance booths to some extent.

A man had to have some kind of stability in his life. His church, his marriage, his home, his club. Any man might need more or less stability than the next. Walters had belonged to four clubs . . . and they were no use to him if he kept meeting Alicia there. And his marriage had broken up, and he wasn't a churchgoer, and a key to his house had been found in Alicia's purse. She should at least have left him his clubs.

Fisher spoke, interrupting his train of thought. "You've been talking about impulse murders, haven't you? Six years of not being able to stand his ex-wife and not being able to get away from her. So finally he hits her with a cigarette box."

"Most of them are impulse murders, yes."

"Well, this wasn't any impulse murder. Look at what he had to do to bring it about. He'd have had to ask her to wait at home for him. Then make some excuse to get away from Larimer, shift home, kill her fast and get back to the Sirius Club before Larimer wonders where he's gone. Then he's got to hope Larimer will forget the whole thing. That's not just cold-blooded, it's also stupid."

"Yah. So far it's worked, though."

"Worked, hell. The only evidence you've got against Walters is that he had good reason to kill her. Listen, if she got on his nerves that much, she may have irritated some other people too."

Hennessey nodded. "That's the problem, all right." But he didn't mean it the way Fisher did.

Walters had moved to a hotel until such time as the police were through with his house. Hennessey called him before going off duty.

"You can move home," he told him.

"That's good," said Walters. "Find out anything?"

"Only that your wife was murdered with that selfsame cigarette box. We found no sign of anyone in the house except her, and you." He paused, but Walters only nodded thoughtfully. He asked, "Did the box look familiar to you?"

"Oh, yes, of course. It's mine. Alicia and I bought it on our honeymoon, in Switzerland. We divided things during the divorce, and that went to me."

"All right. Now, just how violent was that argument you had?"

He flushed. "As usual. I did a lot of shouting, and she just sat there letting it go past her ears. It never did any good."

"Did you strike her?"

The flush deepened, and he nodded. "I've never done that before."

"Did you by any chance hit her with a malachite box?"

"Do I need a lawyer?"

"You're not under arrest, Mr. Walters. But if you feel you need a lawyer, by all means get one." Hennessey hung up.

He had asked to be put on the day shift today, in order to follow up this case. It was quitting time now, but he was reluctant to leave.

Officer-1 Fisher had been eavesdropping. He said, "So?"

"He never mentioned the word *alibi*," said Hennessey. "Smart. He's not supposed to know when she was killed."

"You're still sure he did it."

"Yah. But getting a conviction is something else again. We'll find more people with more motives. And all we've got is the laboratory." He ticked items off on his fingers. "No fingerprints on the box. No blood on Walters or any of his clothes, unless he had paper clothes and ditched 'em. No way of proving Walters let her in or gave her the key . . . though I wonder if he really had that much trouble keeping her out of the house.

"We'd be asking a jury to believe that Walters left the table and Larimer forgot about it. Larimer says no. Walters is pretty sure to get the benefit of the doubt. She didn't bleed much; a good defense lawyer is bound to suggest that she was moved from somewhere else."

"It's possible."

"She wasn't dead until she was hit. Nothing in the stomach but food. No drugs or poisons in the bloodstream. She'd have had to be killed by someone who—" He ticked them off. "Knew she had Walters' key. Knew Walters' displacement booth number. And knew Walters wouldn't be home. Agreed?"

"Maybe. How about Larimer or Lovejoy?"

Hennessey spread his hands in surrender. "It's worth asking. Larimer's alibi is as good as Walters', for all that's worth. And we've still got to interview Jennifer ... Lewis."

"Then again, a lot of people at the Sirius Club knew Walters. Some of them must have been involved with Alicia. Anyone who saw Walters halfway through a domino game would know he'd be stuck there for awhile."

"True. Too true." Hennessey stood up. "Guess I'll be getting dinner."

Hennessey came out of the restaurant feeling pleasantly stuffed and torpid. He turned left toward the nearest booth, a block away.

The Walters case had haunted him all through dinner. Fisher had made a good deal of sense ... but what bugged him was something Fisher hadn't said. Fisher hadn't said that Hennessey might be looking for easy answers.

Easy? If Walters had killed Alicia during a game of dominos at the Sirius Club, then there wouldn't be any case until Larimer remembered. Aside from that, Walters would have been an idiot to try such a thing. Idiot, or desperate.

But if someone else had killed her, it opened up a bag of snakes. Restrict it to members of the Sirius Club who were there that night, and how many were left? They'd both done business there. How many of Jeffrey Walters' acquaintances had shared Alicia's bed? Which one would have killed her, for reason or no reason? The trouble with sharing too many beds was that one's chance of running into a really bad situation was improved almost to certainty.

If Walters had done it, things became simpler.

But she hadn't bled much.

And Walters couldn't have had reason to move her body to his home. Where could he have killed her that would be worse than that?

Walters owned the murder weapon ... no, forget that. She could have been hit with anything, and if she were in Walters' house fifteen seconds later she might still be breathing when the malachite box finished the job.

Hennessey slowed to a stop in front of the booth. Something Fisher *had* said, something that had struck him funny. What was it?

"Her displacement booth must be ten years old—" That was it. The sight of the booth must have sparked that memory. And it *was* funny. How had he known?

JumpShift booths were all alike. They had to be. They all had to hold the same volume, because the air in the receiver had to be flicked back to the transmitter. When JumpShift improved a booth, it was the equipment they improved, so that the older booths could still be used.

Ten years old. Wasn't that—yes! The altitude shift. Pumping energy into a cargo, so that it could be flicked a mile or a hundred miles uphill, had been an early improvement. But a transmitter that could absorb the lost potential energy of a downhill shift, had not become common until ten years ago.

Hennessey stepped in and dialed the police station.

Sergeant Sobel was behind the desk. "Oh, Fisher left an hour ago," he said. "Want his number?"

"Yes . . . No. Get me Alicia Walters' number."

Sobel got it for him. "What's up?"

"Tell you in a minute," said Hennessey, and he flicked out.

It was black night. His ears registered the drop in pressure. His eyes adjusted rapidly, and he saw that there were lights in Alicia Walters' house.

He stepped out of the booth. Whistling, he walked a slow circle around it.

It was a JumpShift booth. What more was there to say? A glass cylinder with a rounded top, big enough for a tall man to stand upright and a meager amount of baggage to stand with him—or for a man holding a dead woman in his arms, clenching his teeth while he tried to free one finger for dialing. The machinery that made the magic was buried beneath the booth. The dial, a simple push-button phone dial. Even the long distance booths looked just like this one, though the auxiliary machinery was far more complex.

"But he was sweating—" Had Lovejoy meant it literally?

Hennessey was smiling ferociously as he stepped back into the booth.

The lights of the Homicide room flashed in his eyes. Hennessey came out tearing at his collar. Sweat started from every pore. Living in the mountains like that, Alicia should certainly have had her booth replaced. The room felt like a furnace, but it was his own body temperature that had jumped seven degrees in a moment. Seven degrees of randomized energy, to compensate for the drop in potential energy between here and Lake Arrowhead.

Walters sat slumped, staring straight ahead of him. "She didn't understand and she didn't care. She was taking it like we'd been all through this before but we had to do it again but let's get it over with." He spoke in a monotone, but the nervous stutter was gone. "Finally I hit her. I guess I was trying to get her attention. She just took it and looked at me and waited for me to go on."

Hennessey said, "Where did the malachite box come in?"

"Where do you think? I hit her with it."

"Then it was hers, not yours."

"It was ours. When we broke up, she took it. Look, I don't want you to think I wanted to *kill* her. I wanted to scar her."

"To scare her?"

"No! To scar her!" His voice rose. "To leave a mark she'd remember every time she looked in a mirror, so she'd know I meant it, so she'd leave me alone! I wouldn't have cared if she sued. Whatever it cost, it would have been worth it. But I hit her too hard, way too hard. I felt the crunch."

"Why didn't you report it?"

"But I did! At least, I tried. I picked her up in my arms and wrestled her out to the booth and dialed for the Los Angeles Emergency Hospital. I don't know if there's anyplace closer, and I wasn't thinking too clear. Listen, maybe I can prove this. Maybe an intern saw me in the booth. I flicked into the Hospital, and suddenly I was broiling. Then I remembered that Alicia had an old booth, the kind that can't absorb a difference in potential energy."

"We guessed that much."

"So I dialed quick and flicked right out again. I had to go back to Alicia's for the malachite box and to wipe off the sofa, and my own booth *is* a new one, so I got the temperature shift again. God, it was hot. I changed suits before I went back to the Club. I was still sweating."

"You thought that raising her temperature would foul up our estimate of when she died."

"That's right." Walters' smile was wan. "Listen, I did try to get her to a hospital. You'll remember that, won't you?"

"Yah. But you changed your mind."

All the Bridges Rusting

Take a point in space.

Take a specific point near the star system Alpha Centaurus, on the line linking the center of mass of that system with Sol. Follow it as it moves toward Sol system at lightspeed. We presume a particle in this point.

Men who deal in the physics of teleportation would speak of it as a "transition particle." But think of it as a kind of super-neutrino. Clearly it must have a rest mass of zero, like a neutrino. Like a neutrino, it must be fearfully difficult to find or stop. Despite several decades in which teleportation has been in common use, nobody has ever directly demonstrated the existence of a "transition particle." It must be taken on faith.

Its internal structure would be fearfully complex in terms of energy states. Its relativistic mass would be twelve thousand two hundred tons.

One more property can be postulated. Its location in space is uncertain: a probability density, thousands of miles across as it passes Proxima Centauri, and spreading. The mass of the tiny red dwarf does not bend its path significantly. As it approaches the solar system the particle may be found anywhere within a vaguely bounded wave front several hundred thousand miles across. This vagueness of position is part of what makes teleportation work. One's aim need not be so accurate.

Near Pluto the particle changes state.

Its relativistic mass converts to rest mass within the receiver cage of a drop ship. Its structure is still fearfully complex for an elementary particle: a twelve-thousand-two-hundred-ton spacecraft, loaded with instruments, its hull windowless and very smoothly contoured. Its presence here is the only evidence that a transition particle ever existed. Within the control cabin, the pilot's finger is still on the TRANSMIT button.

Karin Sagan was short and stocky. Her hands were large; her feet were small and prone to foot trouble. Her face was square and cheerful, her eyes were bright and direct, and her voice was deep for a woman's. She had been thirty-six years old when Phoenix left the transmitter at Pluto. She was three months older now, though nine years had passed on Earth.

She had seen a trace of the elapsed years as Phoenix left the Pluto drop ship. The shuttlecraft that had come to meet them was of a new design, and its attitude jets showed the color of fusion flame. She had wondered how they made fusion motors that small.

She saw more changes now, among the gathered newstapers. Some of the women wore microskirts whose hems were cut at angles. A few of the men wore assymetrical shirts—the left sleeve long, the right sleeve missing entirely. She asked to see one man's left cuff, her attention caught by the glowing red design. Sure enough, it was a functional wristwatch; but the material was soft as cloth.

"It's a Bulova Dali," the man said. He was letting his amusement show. "New to you? Things change in nine years, Doctor."

"I thought they would," she said lightly. "That's part of the fun."

But she remembered the shock of relief when the heat struck. She had pushed the TRANSMIT button a light-month out from Alpha Centaurus B. An instant later sweat was running from every pore of her body.

There had been no guarantee. The probability density that physicists called a transition particle could have gone past the drop ship and out into the universe at large, beyond rescue forever. Or . . . a lot could happen in nine years. The station might have been wrecked or abandoned.

But the heat meant that they had made it. *Phoenix* had lost potential energy entering Sol's gravitational field and had gained it back in heat. The cabin felt like a furnace, but it was their body temperature that had jumped from 98.6° to 102°, all in an instant.

"How was the trip?" The young man asked.

Karin Sagan returned to the present. "Good, but it's good to be back. Are we recording?"

"No. When the press conference starts you'll know it. That's the law. Shall we get it going?"

"Fine." She smiled around the room. It was good to see strange faces again. Three months with three other people in a closed environment . . . it was enough.

The young man led her to a dais. Cameras swiveled to face her, and the conference started.

Q: How was the trip?

"Good. Successful, I should say. We learned everything we wanted to know about the Centaurus systems. In addition, we learned that our systems work. The drop-ship method is feasible. We reached the nearest stars, and we came back, with no ill effects."

Q: What about the Centaurus planets? Are they habitable?

"No." It hurt to say that. She saw the disappointment around her.

Q: Neither of them checked out?

"That's right. There are six known planets circling Alpha Centaurus B. We may have missed a couple that were too small or too far out. We had to do all our looking from a light-month away. We had good hopes for B-2 and B-3— remember, we knew they were there before we set out—but B-2 turns out to be a Venus-type with too much atmosphere, and B-3's got a reducing atmosphere, something like Earth's atmosphere three billion years ago."

Q: The colonists aren't going to like that, are they?

"I don't expect they will. We messaged the drop ship *Lazarus II* to turn off its JumpShift unit for a year. That means that the colony ships won't convert to rest mass when they reach the receiver. They'll be reflected back to the solar system. They should appear in the Pluto drop ship about a month from now."

Q: Having lost nine years.

"That's right. Just like me and the rest of the crew of *Phoenix*. The colonists left the Pluto transmitter two months after we did."

Q: What are the chances of terraforming B-3 someday?

Karin was glad to drop the subject of the colony ships. Somehow she felt that she had failed those first potential colonists of another star system. She said, "Pretty good, someday. I'm just talking off the top of my head, you understand. I imagine it would take thousands of years, and would involve seeding the atmosphere with tailored bacteria and waiting for them to turn methane and ammonia and hydrocarbons into air. At the moment it'll pay us better to go on looking for worlds around other stars. It's so bloody easy, with these interstellar drop ships."

There was nodding among the newstapers. They knew about drop ships, and they had been briefed. In principle there was no difference between *Lazarus II* and the drop ships circling every planet and most of the interesting moons and asteroids in the solar system. A drop ship need not be moving at the same velocity as its cargo. The *Phoenix*, at rest with respect to Sol and the Centaurus suns, had emerged from *Lazarus II*'s receiver cage at a third of lightspeed.

"The point is that you can use a drop ship more than once," Karin went on. "By now *Lazarus II* is one and a third light-years past Centaurus. We burned most of its fuel to get the ship up to speed, but there's still a maneuver reserve. Its next target is an orange-yellow dwarf, Epsilon Indi. *Lazarus II* will be there in about twenty-eight years. Then maybe we'll send another colony group."

Q: Doctor Sagan, you were as far from Sol as anyone in history has ever gotten. What was it like out there?

Karen giggled. "We were as far from any star as anyone's ever gotten. It was a long night. Maybe it was getting to us. We had a bad moment when we thought there was an alien ship coming up behind us." She sobered, for that moment of relief had cost six people dearly. "It turned out to be *Lazarus*.

I'm afraid that's more bad news. Lazarus should have been decelerating. It wasn't. We're afraid something's happened to their drive."

That caused some commotion. It developed that many of the newstapers had never heard of the first Lazarus. Karin started to explain . . . and that turned out to be a mistake.

The first interstellar spacecraft had been launched in 2004, thirty-one years ago.

Lazarus had been ten years in the building; but far more than ten years of labor had gone into her. Her life-support systems ran in a clear line of development back to the first capsules to orbit Earth. The first fusion-electric power plants had much in common with her main drive, and her hydrogen fuel tanks were the result of several decades of trial and error. Liquid hydrogen is tricky stuff. Centuries of medicine had produced suspended-animation treatments that allowed Lazarus to carry six crew members with life-support supplies sufficient for two.

The ship was lovely —at least, her re-entry system was lovely, a swing-wing streamlined exploration vehicle as big as any hypersonic passenger plane. Fully assembled, she looked like a haphazard collection of junk. But she was loved.

There had been displacement booths in 2004: the network of passenger teleportation had already replaced other forms of transportation over most of the world. The cargo ships that lifted Lazarus' components into orbit had been fueled in flight by JumpShift units in the tanks. It was a pity that Lazarus could not take advantage of such a method. But conservation of momentum held. Fuel droplets entering Lazarus's tanks at a seventh of lightspeed would tear them apart.

So Lazarus had left Earth at the end of the Corliss accelerator, an improbably tall tower standing up from a flat asteroid a mile across. The fuel tanks—most of Lazarus's mass—had been launched first. Then the ship itself, with enough maneuvering reserve to run them down. Lazarus had left Earth like a string of toy balloons, and telescopes had watched as she assembled herself in deep space.

She had not been launched into the unknown. The telescopes of Ceres Base had found planets orbiting Alpha Centaurus B. Two of these might be habitable. Failing that, there might at least be seas from which hydrogen could be extracted for a return voyage.

"The first drop ship was launched six years later," Karin told them. "We should have waited. I was five when they launched *Lazarus*, but I've been told that everyone thought that teleportation couldn't possibly be used for space exploration because of velocity differences. If we'd waited we could have put a drop ship receiver cage on *Lazarus* and taken out the life-support system. As it was, we didn't launch *Lazarus II* until—" She stopped to add up dates. "Seventeen years ago. 2018."

Q: Weren't you expecting Lazarus to pass you?

"Not so soon. In fact, we had this timed pretty well. If everything had gone right, the crew of *Lazarus I* would have found a string of colony ships pouring out of *Lazarus II* as it fell across the system. They could have joined up to explore the system, and later joined the colony if that was feasible, or come home on the colony return ship if it wasn't."

Q: As it is, they're in deep shit.

"I'm afraid so. Can you really say that on teevee?"

There were chuckles at her naiveté.

Q: What went wrong? Any idea?

"They gave us a full report with their distress signal. There was some trouble with the plasma pinch effect, and no parts to do a full repair. They tried running it anyway—they didn't have much choice, after all. The plasma stream went wrong and blew away part of the stern. After that there wasn't anything they could do but set up their distress signal and go back into suspended animation."

Q: What are your plans for rescue?

Karin made her second error. "I don't know. We just got back two days ago, and we've spent that time traveling. It's easy enough to pump energy into an incoming transition particle to compensate for a jump in potential energy, but the only drop ship we've got that can *absorb* potential energy is at Mercury. We couldn't just flick in from Pluto; we'd have been broiled. We had to flick in to Earth orbit by way of Mercury, then go down in a shuttlecraft." She closed her eyes to think. "It'll be difficult. By now *Lazarus* must be half a light-year beyond Alpha Centaurus, and *Lazarus II* more than twice that far. We probably can't use *Lazarus II* in a rescue attempt."

Q: Couldn't you drop a receiver cage from Lazarus II, then wait until Lazarus has almost caught up with it?

She smiled indulgently. At least they were asking intelligent questions. "Won't work. *Lazarus II* must have changed course already for Epsilon Indi. Whatever happens is likely to take a long time."

Teevee was mostly news these days. The entertainment programs had been largely taken over by cassettes, which could be sold devoid of advertisements, and which could be aimed at more selective audiences.

And newspapers had died out; but headlines had not. The announcers were saying things like *Centaurus planets devoid of life ... colony ships to return ... failure of Lazarus scout ship engines ... rescue attempts to begin ... details in a moment, but first this word ...*

Jerryberry Jansen of CBA smiled into the cameras. The warmth he felt for his unseen audience was genuine: he regarded himself as a combination of entertainer and teacher, and his approximately twelve million students were the measure of his success. "The Centaurus expedition was by no means a disaster," he told them. "For one thing, the colony fleet—which cost you, the taxpayer, about six hundred and sixty million new dollars nine years ago—can be re-used as is, once the UN Space Authority finds a habitable world. Probably the colonists themselves will not want to wait that long. A new group may have to be retrained.

"As for the interstellar drop ship concept, it works. This has been the first real test, and it went without a hitch. Probably the next use of drop ships will not be a colony expedition at all, but an attempt to rescue the crew of *Lazarus*. The ship was sending its distress signal. There is good reason to think that the crew is still alive.

"Doctor Karin Sagan has pointed out that any rescue attempt will take decades. This is reasonable, in that the distances to be covered are to be measured in light-years. But today's ships are considerably better than *Lazarus* could ever have been."

"You idiot," said Robin Whyte, Ph.D. He twisted a knob with angry force, and the teevee screen went blank. A few minutes later he made two phone calls.

Karin was sightseeing on Earth.

The UN Space Authority had had a new credit card waiting for her, a courtesy she appreciated. Otherwise she would have had to carry a sackful of chocolate dollars for the slots. Her hands quickly fell into the old routine: insert the card, dial, pull it out, and the displacement booth would send her somewhere else.

It was characteristic of Karin that she had not been calling old friends. The impulse was there, and the worn black phone book with its string of nine-year-old names and numbers. But the people she had known must have changed. She was reluctant to face them.

There had been a vindictive impulse to drop in on her ex-husband. *Here I am at thirty-six, and you*— Stupid. Ron knew where she had been for nine years, so why bug the man?

She had cocktails at Mr. A's in San Diego, lunch at Scandia in Los Angeles, and dessert and coffee at Ondine in Sausalito. The sight of the Golden Gate Bridge sparked her to flick in at various booths for various views of all the bridges in the Bay area. For Karin, as for most of humanity, Earth was represented by a small section of the planet.

There had been changes. She got too close to the Bay Bridge and was horrified at the rust. It had never occurred to her that the San Francisco citizenry might let the bridges decay. *Something* could be done with them: line them with shops à la London Bridge, or landscape them over for a park, or run drag races . . . They would make horribly obtrusive corpses. They would ruin the scenery. Still, that had happened before . . .

Some things had not changed. She walked for an hour in King's Free Park, the landscaped section of what had been the San Diego Freeway. The trees had grown a little taller, but the crowds were the same, always different yet always the same. The shops and crowds in the Santa Monica Mall hadn't

changed ... except that the city had filled in the space between the curbs, where people had had to step down into the
empty streets.

She did some shopping in the Mall. To a saleslady in Mag-
nin's West she said, "Dress me." That turned out to be a considerable project, and it cost. When she left, her new clothes
felt odd on her, but they seemed to blend better with the
crowds around her.

She did a lot of flicking around without ever leaving the
booth—the ubiquitous booth that seemed to be one instead of
millions, that seemed to move with her as she explored. It
took her longer to find the right numbers than it did to dial.
But she flicked down the length of Wilshire Boulevard in
jumps of four blocks, from the coast to central Los Angeles,
by simply dialing four digits higher each time.

She stopped off at the Country Art Museum in Fresno and
was intrigued by giant sculptures in plastic foam. She was
wandering through these shapes, just feeling them, not yet
trying to decide whether she liked them, when her wrist
phone rang.

She could have taken the call then and there, but she went
to a wall phone in the lobby. Karin preferred to see who she
was talking to.

She recognized him at once.

Robin Whyte was a round old man, his face pink and soft
and cherubic, his scalp bare but for a fringe of white hair
over his ears and a single tuft at the top of his head. Karin
was surprised to see him now. He was the last living member
of the team that had first demonstrated teleportation in 1992.
He had been president of JumpShift, Inc., for several decades,
but he had retired just after the launching of *Lazarus II*.

"Karin Sagan?" His frown gave him an almost petulant
look. "My congratulations on your safe return."

"Thank you." Karin's smile was sunny. An impulse made
her add, "Congratulations to you, too."

He did not respond in kind. "I need to see you. Urgently.
Can you come immediately?"

"Concerning what?"

"Concerning the interview you gave this morning."

But the interview had gone so well. What could be bothering the man? She said, "All right."

The number he gave her had a New York prefix.

It was evening in New York City. Whyte's apartment was the penthouse floor of a half-empty building. The city itself had lost half its population during the past forty years, and it showed in the walls of dark windows visible through Whyte's picture windows.

"The thing I want to emphasize," said Whyte, "is that I didn't call you here as a representative of JumpShift. I'm retired. But I've got a problem, and pretty quick I'm going to have to take it up with someone in JumpShift. I still own enough JumpShift stock to want to protect it."

His guests made no comment on his disclaimer. They watched as he finished making their drinks and served them. Karin Sagan was curious and a bit truculent at being summoned so abruptly. Jerryberry Jansen had known Whyte too long for that. He was only curious.

"You've put JumpShift in a sticky situation," said Whyte. "Both of you, and the rest of the news media too. Karin, Jerryberry, how do you feel about the space program?"

"I'm for it. You know that," said Jerryberry.

"I'm in it," said Karin. "I feel no strong urge to quit and get an honest job. Is this a preliminary to firing me?"

"No. I do want to know why you went into so much detail on *Lazarus*."

"They asked me. If someone had asked me to keep my mouth shut on the subject I might have. Might not."

"We can't rescue *Lazarus*," said Whyte.

There was an uncomfortable silence. Perhaps it was in both their minds, but it was Jerryberry who said it. "Can't or won't?"

"How long have you known me?"

Jerryberry stopped to count. "Fourteen years, on and off. Look, I'm not saying you'd leave a six-man crew in the lurch if it were feasible to rescue them. But is it *economically* infeasible? Is that it?"

"No. It's impossible." Whyte glared at Karin, who glared back. "You should have figured it out, even if he didn't." He transferred the glare to Jansen. "About that rescue mission you proposed on nationwide teevee. Did you have any details worked out?"

Jerryberry sipped at his Screwdriver. "I'd think it would be obvious. Send a rescue ship. Our ships are infinitely better than anything they had in 2004."

"They're moving at a seventh of lightspeed. What kind of ship could get up the velocity to catch Lazarus and still get back?"

"A drop ship, of course! A drop ship burns all its fuel getting up to speed. Lazarus II is doing a third of lightspeed, and it cost about a quarter of what Lazarus cost—it's so much simpler. You send a drop ship. When it passes Lazarus you drop a rescue ship through."

"Uh huh. And how fast is the rescue ship moving?"

". . . Oh." Lazarus would flash past the rescue ship at a seventh of lightspeed.

"We've got better ships than the best they could do in 2004. Sure we do. But, censored dammit, they don't travel the same way!"

"Well, yes, but there's got to be—"

"You're cheating a little," Karin said. "A rescue ship of the Lazarus type could get up to speed and still have the fuel to get home. Meanwhile you send a drop ship to intercept Lazarus. The rescue ship drops through the receiver cage, picks them up—hmm."

"It would have to be self-teleporting, wouldn't it? Like Phoenix."

"Yah. Hmmm."

"If you put a transmitter hull around something the size of Lazarus, fuel tanks included, you'd pretty near double the weight. It couldn't get up to speed and then decelerate afterward. You'd need more fuel, more weight, a bigger hull. Maybe it couldn't be done at all, but sure as hell we're talking about something a lot bigger than Lazarus."

There had never been another ship as big as Lazarus.

Karin said, "Yah. You'd ditch a lot of fuel tanks getting up to speed, but still—hmm. Fuel to get home. Dammit, Whyte, I left Earth nine years ago. You've had nine years to improve your space industry! What have you done?"

"We've got lots better drop ships," Whyte said quietly. Then, "Don't you understand? We're improving our ships, but not in the direction of a bigger and better *Lazarus*."

Silence.

"Then there's the drop ship itself. We've never built a receiver cage big enough to take another *Lazarus*. *Phoenix* isn't big; it doesn't have to go anywhere. I won't swear it's impossible to build a drop ship that size, but I wouldn't doubt it either. It doesn't matter. We can't build the rescue ship. We don't even have the technology to build *Lazarus* again! It's gone, junked when we started building drop ships!"

"Like those damn big bridges in San Francisco Bay," whispered Karin. "Sorry, gentlemen. I hadn't thought it out."

Jerryberry said, "You've still got the Corliss accelerator. And we still use reaction drives."

"Sure. For interplanetary speeds. And drop ships."

Jerryberry drained his Screwdriver in three swallows. With his mind's eye he saw six coffins, deathly still, and six human beings frozen inside. Three men, three women. Someone must have thought that a scout crew might just decide to colonize the Centaurus system without waiting. Fat chance of that now. Three men, three women, frozen, falling through interplanetary space forever. They couldn't possibly have been expecting rescue. Could they?

"So we don't get them back," he said. "What are we holding, a wake?"

"They knew the risks they were taking," said Whyte. "They knew, and they fought for the chance. We had over a thousand volunteers at the start of training, and that was after the preliminary weeding-out. Jerryberry, I asked you before about how you felt about the space program."

"I told you. In fact—" He stopped. "Publicity."

"Right."

"I thought I was doing you some good. Public support for the space program isn't heavy right now, and frankly, Doctor Sagan, your report didn't help much."

She flared up. "What were we supposed to do, *build* a planet?"

"Failure of the first expedition. No planets. A whole colony fleet on its way home without ever having so much as *seen* Alpha Centaurus! I know, it's safer for them, and better not to waste the time, but dammit!" Jerryberry was on his feet and pacing. There was an odd glow in his eyes, an intensity that could communicate even through a teevee screen. "I tried to emphasize the good points. Now—I damn near promised the world a rescue mission, didn't I?"

"Just about. You weren't the only one."

He paced. "I'm pretty good at explaining. I have to be. I'll just have to tell them—no, let's do it right. Robin, will you go on teevee?"

Whyte looked startled.

"Tell you what," said Jerryberry. "Don't just tell them why we can't rescue *Lazarus*. Show them. Set up a cost breakdown, in dollars and years. We all know—"

"I tell you it isn't *cost*. It—"

"We both know that it *could* be done, if we gave up the rest of the space industry and concentrated solely on rescuing *Lazarus* for enough years. R and D, rebuilding old hardware—"

"Censored dammit! The research alone on a drop ship that size—" Whyte cocked his head as if listening to an inner voice. "That is one way to put it. It would cost us everything we've built up in the past thirty years. Jerryberry, is this really the way to get it across?"

"I don't know. It's one way. Set up a cost estimate you can defend. It won't end with just one broadcast. You'll be challenged, whatever you say. Can you be ready in two days?"

Karin gave a short, barking laugh.

Whyte smiled indulgently. "Are you out of your mind? A valid cost estimate would take months, assuming I can get anyone interested in doing a cost estimate of something nobody really wants built."

Jerryberry paced. "Suppose we do a cost estimate. CBA, I mean. Then you wouldn't have anything to defend. It wouldn't be very accurate, but I'm sure we could get within a factor of two."

"Better give yourselves a week. I'll give you the names of some people at JumpShift; you can go to them for details. Meanwhile I'll have them issue a press release saying we're not planning a rescue mission for *Lazarus* at this time."

JumpShift Experimental Laboratory, Building One, was a tremendous pressurized Quonset hut. On most of his previous visits Jerryberry had found it nearly empty; too many of JumpShift's projects are secret. Once he had come here with a camera team, and on that occasion the polished, smoothly curved hull of *Phoenix* had nearly filled the building.

He had never known exactly where the laboratory was. Its summers and winters matched the Northern Hemisphere, and the sun beyond the windows now stood near noon, which put it on Rocky Mountain time.

Gemini Jones was JumpShift's senior research physicist, an improbably tall and slender black woman made even taller by a head of hair like a great white dandelion. "We get this free," she said, rapping the schematic diagrams spread across the table. "The Corliss accelerator. Robin wants to build another of these. We don't have the money yet. Anyway, we can use it for the initial boost."

On a flattish disk of asteroidal rock a mile across, engineers of the past generation had raised a tower of metal rings. The electromagnetic cannon had been firing ships from Earth orbit since A.D. 2004. Today it was used more than ever, to accelerate the self-transmitting ships partway toward the orbital velocities of Mars, Jupiter, Mercury . . .

Jerryberry studied the tower of rings, wider than any ship ever built. "Is it wide enough for what we've got in mind?"

"I think so. We'd fire the rescue ship in sections, then put it together in space. But we'd still have to put a transmitter hull around it."

"Okay, we've got the accelerator, and we'd use standard tanks. Beyond that—"

"Now hold up," said Gem. "There's an easier way to do this. I thought of it this morning. If we do it my way we won't need any research at all."

"Oh? You interest me strangely."

"See, we've still got this problem of building a ship big enough to make the rescue and then decelerate, and a drop cage big enough to take it. But we already know we can build self-transmitting hulls the size of *Phoenix*. What we can do is put the deceleration fuel in *Phoenix* hulls. We wouldn't need an unreasonably big drop cage that way."

Jerryberry whistled. He knew what *Phoenix* had cost. Putting a rescue ship together would be like building a fleet of *Phoenixes*. And yet—

"Robin was wrong. We could do that. We've got the hardware."

"That's exactly right. I figure maybe twenty *Phoenix* hulls full of slurried hydrogen, plus a *Phoenix*-type ship for the rescue, plus a couple more hulls to hold the drive and the rigging to string it all together. You'd have to assemble it after launch and accelerate it to a seventh of lightspeed, using a couple hundred standard tanks. Then take it apart, stow the rigging, and send everything through a *Lazarus II* drop ship one hull at a time."

"We could do it. Does Robin know about this?"

"Who's had time to call him? I only just thought of this an hour ago. I've been working out the math."

"We could do it," Jerryberry said, his eyes afire. "We could bring 'em back. All it would take would be time and money."

She smiled indulgently down at him; at least she always seemed to, though her eyes were level with his own. "Don't get too involved. Who's going to pay for all this? You might talk your bemused public into it if you were extending man's dominion across the stars. But to rescue six failures?"

"You don't really think of them that way."

"Nope. But somebody's going to say it."

"I don't know. Maybe we should go for it. Those self-transmitting hulls could be turned into ships afterward."

"No. You'd drop them on the way back."

Jerryberry ran a hand through his hair. "I guess you're right. Thanks, Gem. You've done a lot of work for something that isn't ever going to get built."

"Good pactice. Keeps my brain in shape," said Gem.

He was at home, doggedly working out a time-and-costs schedule for the rescue of *Lazarus*, when Karin Sagan called. She said, "I've been wondering if you need me for the broadcast."

"Good idea," said Jerryberry, "if you're willing. We could tape an interview any time you're ready. I'll ask you to describe the circumstances under which you found *Lazarus*, and use that to introduce the topic."

"Good."

Jerryberry was tired and depressed. It took him a moment to see that Karin was too. "What's wrong?"

"Oh . . . a lot of things. We aren't just going to forget about those six astronauts, are we?"

His laugh was brittle. "I think it unlikely. They aren't decently dead. They're in limbo, falling across our sky forever."

"That's what I mean. We could wake them any time in the next thousand years, if we could get to them."

"That's my problem. We can."

"What?"

"But it'd cost the Moon, so to speak. Come on over, Doctor. I'll show you."

Lazarus cost	N$	2,000,000,000

Lazarus II cost	N$	500,000,000
Phoenix cost	N$	110,000,000
Colony (six ships adequately equipped) cost	N$	660,000,000
Support systems in solar system	N$	250,000,000
TOTAL COLONY PACKAGE, IN-CLUDING COLONY AND *PHOENIX* AND SUPPORT SYS-TEMS IN SOLAR SYSTEM:	N$	1,520,000,000

Twenty-two self-transmitting hulls cost (One self-transmitting hull costs N$70,000,000)	N$	1,540,000,000
Interstellar drop ship costs	N$	500,000,000
Phoenix-type rescue ship costs	N$	110,000,000
R & D costs nothing		
Support systems in solar system	N$	250,000,000
TOTAL COST OF RESCUE	N$	2,300,000,000

". . . which is just comfortably more than it cost to build *Lazarus* in the first place, and a lot more than it cost us to not colonize Alpha Centaurus. It wouldn't be impossible to go get them. Just inconvenient and expensive."

"In spades," said Karin. "You'd tie up the Corliss accelerator for a week solid. The whole trip would take about thirty-four years starting from the launching of the drop ship."

"And if it could be done now it could always be done; we couldn't ever forget it until we'd done it. And it would get more difficult every year because *Lazarus* would be getting further away."

"It'll nag us the rest of our lives." Karin leaned back in Jerryberry's guest chair. His apartment was not big: three rooms, with doors knocked between them, in a complex that had been a motel on the Pacific Coast Highway thirty years ago. "There's another thing. What are we really doing if we do it Whyte's way? We're talking the public into not backing a space project. Suppose they got the habit? I don't know about you—"

"I just plain like rocket ships," said Jerryberry.

"Okay. Can you really talk the public into this?"

"No. *Lazarus* didn't even cost this much, and *Lazarus* almost didn't get built, they tell me. And *Lazarus* failed, and so did the colony project. So: no. But I'm not sure I can bring myself to talk them out of it."

"Jansen, just how bad is public support for the Space Authority?"

"Oh ... it isn't even that, exactly. The public is getting unhappy about JumpShift itself."

"What? What for?"

"CBA runs a continuous string of public opinion polls. The displacement booths did genuinely bring some unique problems with them—"

"They solved some too. Maybe you don't remember."

Jerryberry smiled. "I'm not old enough. Neither are you. Slums, traffic jams, plane crashes—nobody's that old except Robin Whyte, and if you try to tell him the booths brought problems of their own, he thinks you're an ungrateful bastard. But they did. You know they did."

"Like flash crowds?"

"Sure. Any time anything interesting happens anywhere, some newstaper is going to report it. Then people flick in to see it from all over the United States. If it gets big enough you get people flicking in just to see the crowd, plus pickpockets, looters, cops, more newstapers, anyone looking for publicity.

"Then there's the drug problem. There's no way to stop smuggling. You can pick a point in the South Pacific with the same longitude and opposite latitude as any given point in the USA and most of Canada, and teleport from there without worrying about the Earth's rotational velocity. All it takes is two booths. You can't stop the drugs from coming in. I remember one narcotics cop telling me to think of it as evolution in action."

"God."

"Oh, and the ecologists don't like the booths. They make wilderness areas too available. And the cops have their problems. A man used to be off the hook if he could prove he was somewhere else when a crime happened. These days you have to suspect anyone, anywhere. The real killer gets lost in the crowd.

"But the real beef is something else. There are people you have to get along with, right?"

"Not me," said Karin.

"Well, you're unusual. Everyone in the world lives next door to his boss, his mother-in-law, the girl he's trying to drop, the guy he's fighting for a promotion. You can't move away from anyone. It bugs people."

"What can they do? Give up the booths?"

"No. There aren't any more cars or planes or railroads. But they can give up space."

Karin thought about that. Presently she gave her considered opinion. "Idiots."

"No. They're just like all of us: they want something for nothing. Have you ever solved a problem without finding another problem just behind it?"

"Sure. My husband . . . well, no, I was pretty lonely after we split up. But I didn't sit down and cry about it. When someone hands me a problem, I solve it. Jansen, we're going at this wrong. I feel it."

"Okay, so we're doing it wrong. What's the right way?"

"I don't know. We've got better ships than anyone dreamed of in 2004. That's fact."

"Define *ship*."

"Ship! Vehicle! Never mind, I see the point. Don't push it."

So he didn't ask her what a 747 circling the sinking Titanic could have done to help, or whether a Greyhound bus could have crossed the continent in 1849. He said, "We know how to rescue *Lazarus*. What's the big decision? We do or we don't."

"Well?"

"I don't know. We watch the opinion polls. I think ∴ I think we'll wind up neutral. Present the project as best we can finagle it up. Tell 'em the easiest way to do it, tell 'em what it'll cost, and leave it at that."

The opinion polls were a sophisticated way to read mass minds. Over the years sampling techniques had improved enormously, raising their accuracy and lowering their cost. Public thinking generally came in blocks:

JumpShift's news release provoked no immediate waves. But one block of thinking began to surface. A significant segment of humanity was old enough to have watched teevee coverage of the launching of *Lazarus*. A smaller, still significant segment had helped to pay for it with their taxes.

It had been the most expensive space project of all time. *Lazarus* had been loved. Nothing but love could have pushed the taxpayer into paying such a price. Even those who had fought the program thirty-one years ago now remembered *Lazarus* with love.

The reaction came mainly from older men and women, but it was worldwide. *Save Lazarus*.

Likewise there were those dedicated to saving the ecology from the intrusion of Man. For them the battle was never-ending. True, industrial wastes were no longer dumped into the air and water; the worst of these were flicked through a drop ship in close orbit around Venus, to disappear into the atmosphere of that otherwise useless world. But the ultimate garbage-maker was himself the most dangerous of threats. Hardly a wilderness was left on Earth that was not being settled by men with JumpShift booths.

They would have fought JumpShift on any level. JumpShift proposed to leave three men and three women falling across the sky forever. To hell with their profit margin: save *Lazarus*.

There were groups who would vote against anything done in space. The returns from space exploration had been great, admittedly, but they all derived from satellites in close orbit around Earth: observatories, weather satellites, teevee transmitters, solar power plants. These were dirt cheap these days, and their utility had surely been obvious to any moron since Neanderthal times.

But what use were the worlds of other stars? Even the worlds of the solar system had given no benefit to Man, except for Venus, which made an excellent garbage dump. Better to spend the money on Earth. *Abandon Lazarus*.

But most of the public voted a straight *Insufficient Data*. And of course they were right.

Robin Whyte was nervous. He was trying not to show it, but he paced too much and he smiled too much and he kept clasping his hands behind his back. "Sit down, for Christ's sake," said Jerryberry. "Relax. They can't throw tomatoes through their teevee screens."

Whyte laughed. "We're working on that in the research division. Are you almost ready?"

"An hour to broadcast. I've already done the interview with Doctor Sagan. It's on tape."

"Let's see what you've got."

What CBA had for this broadcast was a fully detailed rescue project, complete with artist's conceptions. Jerryberry spread the paintings along a wall. "Using your artists, whom we hired for a week with JumpShift's kind permission. Aren't they beautiful? We also have a definite price tag. Two billion three hundred million new dollars."

Whyte's laugh was still shaky. "That's right on the borderline. Barely feasible." He was looking at an artist's conception of the launching of the rescue mission: a stream of spherical fuel tanks and larger, shark-shaped Phoenix hulls pouring up through the ringed tower of the Corliss accelerator. More components rested on flat rock at the launch end. "So Gem thought of it first. I must be getting old."

"You don't expect to think of everything, do you? You once told me that your secretary thought of the fresh-water tower gimmick during a drunken office party."

"True, too. I paid her salary for thirty years, hoping she'd do it again, but she never did ... Do you think they'll buy it?"

"No."

"I guess not." Whyte seemed to shake himself. "Well, maybe we'll use it some other time. It's a useful technique, shipping fuel in Phoenix hulls. We'll probably need it to explore, say, Barnard's Star, which is moving pretty censored fast with respect to Sol."

"We don't have to tell them they can't do it. Just tell 'em the price tag and let them make up their own minds."

"Listen, I had a hand in launching Lazarus. The launching boosters were fueled by JumpShift units."

"I know."

Whyte, prowling restlessly, was back in front of the launching scene. "I always thought they should have drilled right through the asteroid. Leave the Corliss accelerator open at both ends."

Activity in the sound studio had diminished. Against a white wall men had placed a small table and two chairs, and a battery of teevee cameras and lights were aiming their muzzles into the scene.

Jerryberry touched Whyte's arm. "Let's go sit down over there." Whyte might freeze up if confronted by the cameras too suddenly. Give him a chance to get used to it.

Whyte didn't move. His head was cocked to one side, and his lips moved silently.

"What's the matter?"

Whyte made a shushing motion.

Jerryberry waited.

Presently Whyte looked up. "You'll have to scrap this. How much time have we got?"

"But— An hour. Less. What do you mean, scrap it?"

Whyte smiled. "I just thought of something. Get me to a telephone, will you? Has Gem still got the schematics of the Corliss accelerator?"

An hour to broadcast time, and Jerryberry began to shake. "Robin, are we going to have a broadcast or not?"

Whyte patted him on the arm. "Count on it."

Gem Jones's big white-on-blue schematic had been thumbtacked to the white wall over the table and chairs. Below it, Jerryberry Jansen leaned back, seemingly relaxed, watching Whyte move about with a piece of chalk.

A thumbtacked blueprint and a piece of chalk. It was slipshod by professional standards. Robin Whyte had not appeared on teevee in a couple of decades. He made professional mistakes: he turned his back on the audience, he covered what he was drawing with the chalk. But he didn't look nervous. He grinned into the cameras as if he could see old friends out there.

"The heart of it is the Corliss accelerator," he said, and with the chalk he drew an arc underneath the tower's launch cradle, through the rock itself. "We excavate here, carve out a space to get the room. Then—" He drew it in.

A JumpShift drop ship receiver cage.

"The rescue ship is self-transmitting, of course. As it leaves the accelerator it transmits back to the launch end. What we have then is an electromagnetic cannon of infinite length. We spin it on its axis so it doesn't get out of alignment. We give the ship an acceleration of one gee for a bit less than two months to boost it to the velocity of *Lazarus*, then we flick it out to the drop ship.

"This turns out to be a relatively cheap operation," Whyte said. "We could put some extra couches in *Phoenix* and use that. We could even use the accelerator to boost the drop ship up to speed, but that would take four months, and we'd have to do it *now*. It would mean building another Corliss accelerator, but—" Whyte grinned into the cameras. "—we should have done that anyway, years ago. There's enough traffic to justify it.

"Return voyage is just as simple. After they pick up the crew of *Lazarus*, they flick to the Pluto drop ship, which is big enough to catch them, then to the Mercury drop ship to lose their potential energy, then back to the Corliss accelerator drop cage. We use the accelerator for another two months to slow it down. The cost of an interstellar drop ship is half a billion new dollars. A new Corliss accelerator would cost us about the same, and we can use it commercially. Total price is half of what *Lazarus* cost." Whyte put down the chalk and sat.

Jerryberry said, "When can you go ahead with this, Doctor?"

"JumpShift will submit a time-and-costs schedule to the UN Space Authority. I expect it'll go to the world vote."

"Thank you, Doctor Whyte, for ..." It was a formula. When the cameras were off Jerryberry sagged in his chair. "Now I can say it. Boy, are you out of practice."

"What do you mean? Didn't I get it across?"

"I think you did. I hope so. You smiled a lot too much. On camera that makes you look self-satisfied."

"I know, you told me before," said Whyte. "I couldn't help it. I just felt so *good*."

There Is a Tide

I

Then, the planet had no name. It circles a star which in 2830 lay beyond the fringe of known space, a distance of nearly forty light-years from Sol. The star is a G3, somewhat redder than Sol, somewhat smaller. The planet, swinging eighty million miles from its primary in a reasonably circular orbit, is a trifle cold for human tastes.

In the year 2830 one Louis Gridley Wu happened to be passing. The emphasis on accident is intended. In a universe the size of ours almost anything that can happen, will. Take the coincidence of his meeting—

But we'll get to that.

Louis Wu was one hundred and eighty years old. As a regular user of boosterspice, he didn't show his years. If he didn't get bored first, or broke, he might reach a thousand.

"But," he sometimes told himself, "not if I have to put up with any more cocktail parties, or bandersnatch hunts, or painted flatlanders swarming through an anarchy park too small for them by a factor of ten. Not if I have to live through another one-night love affair or another twenty-year marriage or another twenty-minute wait for a transfer booth that blows its zap just as it's my turn. And people. Not if I have to live with people, day and night, all those endless centuries."

When he started to feel like that, he left. It had happened three times in his life, and now a fourth. Presumably, it would keep happening. In such a state of anomie, of acute anti-everything, he was no good to anyone, especially his friends, most especially himself. So he left. In a small but adequate spacecraft, his own, he left everything and everyone, heading outward for the fringe of known space. He would not return until he was desperate for the sight of a human face, the sound of a human voice.

On the second trip he had gritted his teeth and waited until he was desperate for the sight of a kzinti face.

That was a long trip, he remembered. And, because he had only been three and a half months in space on this fourth trip, and because his teeth still snapped together at the mere memory of a certain human voice ... because of these things, he added, "I think this time I'll wait till I'm desperate to see a kdatlyno. Female, of course."

Few of his friends guessed the wear and tear these trips saved him. And them. He spent the months reading, while his library played orchestrated music. By now he was well clear of known space. Now he turned the ship ninety degrees, beginning a wide circular arc with Sol at its center.

He approached a certain G3 star. He dropped out of hyperdrive well clear of the singularity in hyperspace which surrounds any large mass. He accelerated into the system on his main thruster, sweeping the space ahead of him with the deep-radar. He was not looking for habitable planets. He was looking for Slaver stasis boxes.

If the pulse returned no echo, he would accelerate until he could shift to hyperdrive. The velocity would stay with him, and he could use it to coast through the next system he tried, and the next, and the next. It saved fuel.

He had never found a Slaver stasis box. It did not stop him from looking.

As he passed through the system, the deep-radar showed him planets like pale ghosts, light gray circles on the white screen. The G3 sun was a wide gray disk, darkening almost to black at the center. The near-black was degenerate matter, compressed past the point where electron orbits collapse entirely.

He was well past the sun, and still accelerating when the screen showed a tiny black fleck.

"No system is perfect, of course," he muttered as he turned off the drive. He talked to himself a good deal, out here where nobody could interrupt him.

"It usually saves fuel," he told himself a week later. By then he was out of the singularity, in clear space. He took the ship into hyperdrive, circled halfway round the system, and began decelerating. The velocity he'd built up during those first two weeks gradually left him. Somewhere near where he'd found a black speck in the deep-radar projection, he slowed to a stop.

Though he had never realized it until now, his system for saving fuel was based on the assumption that he would never find a Slaver box. But the fleck was there again, a black dot on the gray ghost of a planet. Louis Wu moved in.

The world looked something like Earth. It was nearly the same size, very much the same shape, somewhat the same color. There was no moon.

Louis used his telescope on the planet and whistled appreciatively. Shredded white cloud over misty blue . . . faint continental outlines . . . a hurricane whorl near the equator. The ice caps looked big, but there would be warm climate near the equator. The air looked sweet and noncarcinogenic on the spectrograph. And nobody on it. Not a soul!

No next door neighbors. No voices. No faces.

"What the hell," he chortled. "I've got my box. I'll just spend the rest of my vacation here. No men. No women. No children." He frowned and rubbed the fringe of hair along his jaw. "Am I being hasty? Maybe I should knock."

But he scanned the radio bands and got nothing. Any civilized planet radiates like a small star in the radio range. Moreover, there was no sign of civilization, even from a hundred miles up.

"Great! Okay, first I'll get that old stasis box." He was sure it was that. Nothing but stars and stasis boxes were dense enough to show black in the reflection of a hyperwave pulse.

He followed the image around the bulge of the planet. It

seemed the planet had a moon after all. The moon was twelve hundred miles up, and it was ten feet across.

"Now why," he wondered aloud, "would the Slavers have put it in orbit? It's too easy to find. They were in a war, for Finagle's sake! And why would it stay here?"

The little moon was still a couple of thousand miles away, invisible to the naked eye. The scope showed it clearly enough. A silver sphere ten feet through, with no marks on it.

"A billion and a half years it's been there," said Louis to himself, said he. "And if you believe that, you'll believe anything. Something would have knocked it down. Dust, a meteor, the solar wind. Tnuctip soldiers. A magnetic storm. Nah." He ran his fingers through straight black hair grown too long. "It must have drifted in from somewhere else. Recently. Wha—"

Another ship, small and conical, had appeared behind the silvery sphere. Its hull was green, with darker green markings.

II

"Damn," said Louis. He didn't recognize the make. It was no human ship. "Well, it could be worse. They could have been people." He used the com laser.

The other ship braked to a stop. In courtesy, so did Louis.

"Would you believe it?" he demanded of himself. "Three years total time I've spent looking for stasis boxes. I finally find one, and now something else wants it too!"

The bright blue spark of another laser glowed in the tip of the alien cone. Louis listened to the autopilot-computer chuckling to itself as it tried to untangle the signals in an unknown laser beam. At least they did use lasers, not telepathy or tentacle-waving or rapid changes in skin color.

A face appeared on Louis's screen.

It was not the first alien he had seen. This, like some others, had a recognizable head: a cluster of sense organs grouped around a mouth, with room for a brain. Trinocular vision, he noted; the eyes set deep in sockets, well protected, but restricted in range of vision. Triangular mouth, too, with yellow, serrated bone knives showing their edges behind three gristle lips.

Definitely, this was an unknown species.

"Boy, are you ugly," Louis refrained from saying. The alien's translator might be working by now.

His own autopilot finished translating the alien's first message. It said, "Go away. This object belongs to me."

"Remarkable," Louis sent back. "Are you a Slaver?" The being did not in the least resemble a Slaver, and the Slavers had been extinct for eons.

"That word was not translated," said the alien. "I reached the artifact before you did. I will fight to keep it."

Louis scratched at his chin, at two week's growth of bristly beard. His ship had very little to fight with. Even the fusion plant which powered the thruster was designed with safety in mind. A laser battle, fought with com lasers turned to maximum, would be a mere endurance test; and he'd lose, for the alien ship had more mass to absorb more heat. He had no weapons per se. Presumably the alien did.

But the stasis box was a big one.

The Tnuctipun-Slaver war had wiped out most of the intelligent species of the galaxy, a billion and a half years ago. Countless minor battles must have occurred before a Slaver-developed final weapon was used. Often the Slavers, losing a battle, had stored valuables in a stasis box, and hidden it against the day they would again be of use.

No time passed inside a closed stasis box. Alien meat a billion and a half years old had emerged still fresh from its hiding place. Weapons and tools showed no trace of rust. Once a stasis box had disgorged a small, tarsierlike sentient being, still alive. That former slave had lived a strange life before the aging process claimed her, the last of her species.

Slaver stasis boxes were beyond value. It was known that the tnuctipun, at least, had known the secret of direct conversion of matter. Perhaps their enemies had too. Someday, in a stasis box somewhere outside known space, such a device would be found. Then fusion power would be as obsolete as internal combustion.

And this, a sphere ten feet in diameter, must be the largest stasis box ever found.

"I too will fight to keep the artifact," said Louis. "But consider this. Our species has met once, and will meet again regardless of who takes the artifact now. We can be friends or enemies. Why should we risk this relationship by killing?"

The alien sense-cluster gave away nothing. "What do you propose?"

"A game of chance, with the risks even on both sides. Do you play games of chance?"

"Emphatically yes. The process of living is a game of chance. To avoid chance is insanity."

"That it is. Hmmm." Louis regarded the alien head that seemed to be all triangles. He saw it abruptly whip around, flick, to face straight backward, and snap back in the same instant. The sight did something to the pit of his stomach.

"Did you speak?" the alien asked.

"No. Won't you break your neck that way?"

"Your question is interesting. Later we must discuss anatomy. I have a proposal."

"Fine."

"We shall land on the world below us. We will meet between our ships. I will do you the courtesy of emerging first. Can you bring your translator?"

He could connect the computer with his suit radio. "Yes."

"We will meet between our ships and play some simple game, familiar to neither of us, depending solely on chance. Agreed?"

"Provisionally. What game?"

The image on the screen rippled with diagonal lines. Something interfering with the signal? It cleared quickly. "There is a mathematics game," said the alien. "Our mathematics will certainly be similar."

"True." Though Louis had heard of some decidedly peculiar twists in alien mathematics.

"The game involves a screee—" Some word that the autopilot couldn't translate. The alien raised a three-clawed hand, holding a lens-shaped object. The alien's mutually opposed fingers turned it so that Louis could see the different markings on each side. "This is a screee. You and I will throw it upward six times each. I will choose one of the symbols, you will choose the other. If my symbol falls looking upward more often than yours, the artifact is mine. The risks are even."

The image rippled, then cleared.

"Agreed," said Louis. He was a bit disappointed in the simplicity of the game.

"We shall both accelerate away from the artifact. Will you follow me down?"

"I will," said Louis.

The image disappeared.

III

Louis Wu scratched at a week's growth of beard. What a way to greet an alien ambassador! In the worlds of men Louis Wu dressed impeccably; but out here he felt free to look like death warmed over, all the time.

But how was a—Trinoc supposed to know that he should have shaved? No, that wasn't the problem.

Was he fool or genius?

He had friends, many of them, with habits like his own. Two had disappeared decades ago; he no longer remembered their names. He remembered only that each had gone hunting stasis boxes in this direction and that each had neglected to come back.

Had they met alien ships?

There were any number of other explanations. Half a year or more spent alone in a single ship was a good way to find out whether you liked yourself. If you decided you didn't, there was no point in returning to the worlds of men.

But there were aliens out here. Armed. One rested in orbit five hundred miles ahead of his ship, with a valuable artifact halfway between.

Still, gambling was safer than fighting. Louis Wu waited for the alien's next move.

That move was to drop like a rock. The alien ship must have used at least twenty gees of push. After a moment of shock, Louis followed under the same acceleration, protected by his cabin gravity. Was the alien testing his maneuverability?

Possibly not. He seemed contemptuous of tricks. Louis, trailing the alien at a goodly distance, was now much closer to the silver sphere. Suppose he just turned ship, ran for the arti-fact, strapped it to his hull and kept running?

Actually, that wouldn't work. He'd have to slow to reach the sphere; the alien wouldn't have to slow to attack. Twenty gees was close to his ship's limit.

Running might not be a bad idea, though. What guarantee had he of the alien's good faith? What if the alien "cheated"?

That risk could be minimized. His pressure suit had sensors to monitor his body functions. Louis set the autopilot to blow the fusion plant if his heart stopped. He rigged a signal button on his suit to blow the plant manually.

The alien ship burned bright orange as it hit air. It fell free and then slowed suddenly, a mile over the ocean. "Show-off," Louis muttered and prepared to imitate the maneuver.

The conical ship showed no exhaust. Its drive must be either reactionless drive, like his own, or a kzin-style induced gravity drive. Both were neat and clean, silent, safe to bystanders and highly advanced.

Islands were scattered across the ocean. The alien circled, chose one at seeming random, and landed like a feather along a bare shoreline.

Louis followed him down. There was a bad moment while he waited for some unimaginable weapon to fire from the grounded ship, to tear him flaming from the sky while his attention was distracted by landing procedures. But he landed without a jar, several hundred yards from the alien cone.

"An explosion will destroy both our ships if I am harmed," he told the alien via signal beam.

"Our species seem to think alike. I will now descend."

Louis watched him appear near the nose of the ship, in a wide circular airlock. He watched the alien drift gently to the sand. Then he clamped his helmet down and entered the air-lock.

Had he made the right decision?

Gambling was safer than war. More fun, too. Best of all, it gave him better odds.

"But I'd hate to go home without that box," he thought. In nearly two hundred years of life, he had never done anything as important as finding a stasis box. He had made no discoveries, won no elective offices, overthrown no governments. This was his big chance.

"Even odds," he said, and turned on the intercom as he descended.

His muscles and semicircular canals registered about a gee. A hundred feet away waves slid hissing up onto pure white sand. The waves were green and huge, perfect for riding; the beach a definite beer party beach.

Later, perhaps he would ride those waves to shore on his belly, if the air checked out and the water was free of predators. He hadn't had time to give the planet a thorough checkup.

Sand tugged at his boots as he went to meet the alien.

The alien was five feet tall. He had looked much taller descending from his ship, but that was because he was mostly leg. More than three feet of skinny leg, a torso like a beer barrel, and no neck. Impossible that his neckless neck should be so supple. But the chrome yellow skin fell in thick rolls around the bottom of his head, hiding anatomical details.

His suit was transparent, a roughly alien-shaped balloon, constricted at the shoulder, above and below the complicated elbow joint, at the wrist, at hip and knee. Air jets showed at wrist and ankle. Tools hung in loops at the chest. A back pack hung from the neck, under the suit. Louis noted all these tools with trepidation; any one of them could be a weapon.

"I expected that you would be taller," said the alien.

"A laser screen doesn't tell much, does it? I think my translator may have mixed up right with left, too. Do you have the coin?"

"The *screee*?" The alien produced it. "Shall there be no preliminary talk? My name is *screee*."

"My machine can't translate that. Or pronounce it. My name is Louis. Has your species met others besides mine?"

"Yes, two. But I am not an expert in that field of knowledge."

"Nor am I. Let's leave the politenesses to the experts. We're here to gamble."

"Choose your symbol," said the alien, and handed him the coin.

Louis looked it over. It was a lens of platinum or some-thing similar, sharp-edged, with the three-clawed hand of his new gambling partner stamped on one side and a planet, with heavy ice caps outlined, decorating the other. Maybe they weren't ice caps, but continents.

He held the coin as if trying to choose. Stalling. Those gas jets seemed to be attitude jets, but maybe not. Suppose he won? Would he win only the chance to be murdered?

But they'd both die if his heart stopped. No alien could have guessed what kind of weapon would render him helpless without killing him.

"I choose the planet. You flip first."

The alien tossed the coin in the direction of Louis's ship. Louis's eyes followed it down, and he took two steps to re-trieve it. The alien stood beside him when he rose.

"Hand," he said. "My turn." He was one down. He tossed the coin. As it spun cleaming, he saw for the first time that the alien ship was gone.

"What gives?" he demanded.

"There's no need for us to die," said the alien. It held something that had hung in a loop from its chest. "This is a weapon, but both will die if I use it. Please do not try to reach your ship."

Louis touched the button that would blow his power plant.

"My ship lifted when you turned your head to follow the screee. By now my ship is beyond range of any possible ex-plosion you can bring to bear. There is no need for us to die, provided you do not try to reach your ship."

"Wrong. I can leave your ship without a pilot." He left his hand where it was. Rather than be cheated by an alien in a gambling game—

"The pilot is still on board, with the astrogator and the screee. I am only the communications officer. Why did you as-sume I was alone?"

Louis sighed and let his arm fall. "Because I'm stupid," he said bitterly. "Because you used the singular pronoun, or my computer did. Because I thought you were a gambler."

"I gambled that you would not see my ship take off, that you would be distracted by the coin, that you could see only from the front of your head. The risks seemed better than one-half."

Louis nodded. It all seemed clear.

"There was also the chance that you had lured me down to destroy me." The computer was still translating into the first person singular. "I have lost at least one exploring ship that flew in this direction."

"Not guilty. So have we." A thought struck him, and he said, "Prove that you hold a weapon."

The alien obliged. No beam showed, but sand exploded to Louis's left, with a vicious *crack!* and a flash the color of lightning. The alien held something that made holes.

So much for that. Louis bent and picked up the coin. "As long as we're here, shall we finish the game?"

"To what purpose?"

"To see who would have won. Doesn't your species gamble for pleasure?"

"To what purpose? We gamble for survival."

"Then Finagle take your whole breed!" he snarled and flung himself to the sand. His chance for glory was gone, tricked away from him. There is a tide that governs men's affairs ... and there went the ebb, carrying statues to Louis Wu, history books naming Louis Wu, jetsam on the tide.

"Your attitude is puzzling. One gambles only when gambling is necessary."

"Nuts."

"My translator will not translate that comment."

"Do you know what that artifact is?"

"I know of the species who built that artifact. They traveled far."

"We've never found a stasis box that big. It must be a vault of some kind."

"It is thought that that species used a single weapon to end their war and all its participants."

The two looked at each other. Possibly each was thinking the same thing. *What a disaster, if any but my own species should take this ultimate weapon!*

But that was anthropomorphic thinking. Louis knew that a kzin would have been saying: *Now I can conquer the universe, as is my right.*

"Finagle take my luck!" said Louis Wu between his teeth. "Why did you have to show at the same time I did?"

"That was not entirely chance. My instruments found your craft as you backed into the system. To reach the vicinity of the artifact in time, it was necessary to use thrust that damaged my ship and killed one of my crew. I earned possession of the artifact."

"By cheating, damn you!" Louis stood up . . .

And something meshed between his brain and his semicircular canals.

IV

One gravity.

The density of a planet's atmosphere depended on its gravity, and on its moon. A big moon would skim away most of the atmosphere, over the billions of years of a world's evolution. A moonless world the size and mass of Earth should have unbreathable air, impossibly dense, worse than Venus.

But this planet had no moon. Except—

The alien said something, a startled ejaculation that the computer refused to translate. "*Screee!* Where did the water go?"

Louis looked. What he saw puzzled him only a moment. The ocean had receded, slipped imperceptibly away, until what showed now was half a mile of level, slickly shining sea bottom.

"Where did the water go? I do not understand."

"I do."

"Where did it go? Without a moon, there can be no tides. Tides are not this quick in any case. Explain, please."

"It'll be easier if we use the telescope in my ship."

"In your ship there may be weapons."

"Now pay attention," said Louis. "Your ship is very close to total destruction. Nothing can save your crew but the com laser in my ship."

The alien dithered, then capitulated. "If you have weapons, you would have used them earlier. You cannot stop my ship now. Let us enter your ship. Remember that I hold my weapon."

The alien stood beside him in the small cabin, his mouth working disturbingly around the serrated edges of his teeth as Louis activated the scope and screen. Shortly a starfield appeared. So did a conical spacecraft, painted green with darker green markings. Along the bottom of the screen was the blur of thick atmosphere.

"You see? The artifact must be nearly to the horizon. It moves fast."

"That fact is obvious even to low intelligence."

"Yah. Is it obvious to you that this world must have a massive satellite?"

"But it does not, unless the satellite is invisible."

"Not invisible. Just too small to notice. But then, it must be very dense."

The alien didn't answer.

"Why did we assume the sphere was a Slaver stasis box? Its shape was wrong; its size was wrong. But it was shiny, like the surface of a stasis field, and spherical, like an artifact. Planets are spheres too, but gravity wouldn't ordinarily pull something ten feet wide into a sphere. Either it would have to be very fluid, or it would have to be very dense. Do you understand me?"

"No."

"I don't know how your equipment works. My deep-radar uses a hyperwave pulse to find stasis boxes. When something stops a hyperwave pulse, it's either a stasis box, or it's something denser than degenerate matter, the matter inside a normal star. And this object is dense enough to cause tides."

A tiny silver bead had drifted into view ahead of the cone. Telescopic foreshortening seemed to bring it right alongside the ship. Louis reached to scratch at his beard and was stopped by his faceplate.

"I believe I understand you. But how could it happen?"

"That's guesswork. Well?"

"Call my ship. They would be killed. We must save them!"

"I had to be sure you wouldn't stop me." Louis Wu went to work. Presently a light glowed; the computer had found the alien ship with its com laser.

He spoke without preliminaries. "You must leave the spherical object immediately. It is not an artifact. It is ten feet of nearly solid neutronium, probably torn loose from a neutron star."

There was no answer, of course. The alien stood behind him but did not speak. Probably his own ship's computer could not have handled the double translation. But the alien was making one two-armed gesture, over and over.

The green cone swung sharply around, broadside to the telescope.

"Good, they're firing lateral," said Louis to himself. "Maybe they can do a hyperbolic past it." He raised his voice. "Use all the power available. You *must* pull away."

The two objects seemed to be pulling apart. Louis suspected that that was illusion, for the two objects were almost in line-of-sight. "Don't let the small mass fool you," he said, unnecessarily now. "Computer, what's the mass of a ten-foot neutronium sphere?—Around two times ten to the minus six times the mass of this world, which is pretty tiny, but if you get too close . . . Computer, what's the surface gravity?—I don't *believe* it."

The two objects seemed to be pulling together again. Damn, thought Louis. If they hadn't come along, that'd be me.

He kept talking. It wouldn't matter now, except to relieve his own tension. "My computer says ten million gravities at the surface. That may be off. Newton's formula for gravity. Can you hear me?"

"They are too close," said the alien. "By now it is too late to save their lives." It was happening as he spoke. The ship began to crumble a fraction of a second before impact. Impact looked no more dangerous than a cannonball striking the wall of a fort. The tiny silver bead simply swept through the side of the ship. But the ship closed instantly, all in a moment, like tinsel paper in a strong man's fist. Closed into a bead glowing yellow with heat. A tiny sphere ten feet through or a bit more.

"I mourn," said the alien.

"Now I get it," said Louis. "I wondered what was fouling

our laser messages. That chunk of neutronium was right between our ships, bending the light beams."

"Why was this trap set for us?" cried the alien. "Have we enemies so powerful that they can play with such masses?"

A touch of paranoia? Louis wondered. Maybe the whole species had it. "Just a touch of coincidence. A smashed neutron star."

For a time the alien did not speak. The telescope, for want of a better target, remained focused on the bead. Its glow had died.

The alien said, "My pressure suit will not keep me alive long."

"We'll make a run for it. I can reach Margrave in a couple of weeks. If you can hold out that long, we'll set up a tailored environment box to hold you until we think of something better. It only takes a couple of hours to set one up. I'll call ahead."

The alien's triple gaze converged on him. "Can you send messages faster than light?"

"Sure."

"You have knowledge worth trading for. I'll come with you."

"Thanks a whole lot." Louis Wu started punching buttons. "Margrave. Civilization. People. Faces. Voices. Bah." The ship leapt upward, ripping atmosphere apart. Cabin gravity wavered a little, then settled down.

"Well," he told himself. "I can always come back."

"You will return here?"

"I think so," he decided.

"I hope you will be armed."

"What? More paranoia?"

"Your species is insufficiently suspicious," said the alien. "I wonder that you have survived. Consider this neutronium object as a defense. Its mass pulls anything that touches it into a smooth and reflective spherical surface. Should any vehicle approach this world, its crew would find this object quickly. They would assume it is an artifact. What other assumption could they make? They would draw alongside for a closer examination."

"True enough, but that planet's empty. Nobody to defend."

"Perhaps."

The planet was dwindling below. Louis Wu swung his ship toward deep space.

Bigger Than Worlds

Just because you've spent all your life on one planet, doesn't mean that everyone always will. Already there are alternatives to worlds. The Apollo spacecraft have an excellent record; they have never killed anyone in space. The Soviet space station may have killed its inhabitants, but the American Skylab didn't.

Alas, they all lack a certain something. Gravity. Permanence. We want something to live on, or in, something superior to what we've got: safer, or more mobile, or roomier. Otherwise, why move?

It's odd how much there is to be said about structures larger than worlds, considering that we cannot yet begin to build any one of them. On the basis of size, the Dyson sphere—a spherical shell around a sun—comes about in the middle. But let's start small and work our way up.

The Multi-Generation Ship

Robert Heinlein's early story "Universe" has been imitated countless times by most of the writers in the business.

The idea was this: Present-day physics poses a limit on the speed of an interstellar vehicle. The ships we send to distant stars will be on one-way journeys, at least at first. They will have to carry a complete ecology; they couldn't carry enough food and oxygen in tanks. Because they will take generations to complete their journeys, they must also carry a viable and complete society.

Clearly we're talking about quite a large ship, with a population in the hundreds at least: high enough to prevent genetic drift. Centrifugal force substitutes for gravity. We're going to be doing a lot of that. We spin the ship on its axis, and put all the things that need full gravity at the outside, along the hull. Plant rooms, exercise rooms, et cetera. Things that don't need gravity, like fuel and guidance instruments, we line along the axis. If our motors thrust through the same axis, we will have to build a lot of the machinery on tracks, because the aft wall will be the floor when the ship is under power.

The "Universe" ship is basic to a discussion of life in space. We'll be talking about much larger structures, but they are designed to do the same things on a larger scale: to provide a place to live, with as much security and variety and pleasure as Earth itself offers—or more.

Gravity

Gravity is basic to our lifestyle. It may or may not be necessary to life itself, but we'll want it if we can get it, whatever we build.

I know of only four methods of generating gravity aboard spacecraft.

Centrifugal force looks much the most likely. There is a drawback: coriolis effects would force us to re-learn how to walk, sit down, pour coffee, throw a baseball. But its effects would decrease with increasing moment arm, that is, with larger structures. On the Ring City you'd never notice it.

Our second choice is to use actual mass: plate the floor with neutronium, for instance at a density of fifty quadrillion tons per cubic foot; or build the ship around a quantum black hole, invisibly small and around as massive as, say, Phobos. But this will vastly increase our fuel consumption if we expect the vehicle to go anywhere.

Third choice is to generate gravity waves. This may remain forever beyond our abilities. But it's one of those things that people are going to keep trying to build forever, because it would be so damn useful. We could launch ships at a million gravities, and the passengers would never feel it. We could put laboratories on the sun, or colonize Jupiter. Anything.

The fourth method is to accelerate all the way, making turnover at the midpoint and decelerating the rest of the way. This works fine. Over interstellar distances it would take an infinite fuel supply—and by God we may have it, in the Bussard ramjet. A Bussard ramjet would use an electromagnetic field to scoop up the interstellar hydrogen ahead of it—with an intake a thousand miles or more in diameter—compress it, and burn it as fuel for a fusion drive. Now the multi-generation ship would become unnecessary as relativity shortens our trip time: four years to the nearest star, twenty-one years to the galactic hub, twenty-eight to Andromeda galaxy—all at one gravity acceleration.

The Bussard ramjet looks unlikely. It's another ultimate, like generated gravity. Is the interstellar medium sufficiently ionized for such finicky control? Maybe not. But it's worth a try.

Meanwhile, our first step to other worlds is the "Universe" ship—huge, spun for gravity, its population in the hundreds, its travel time in generations.

Flying Cities

James Blish used a variant of generated gravity in his tales of the Okie cities.

His "spindizzy" motors used a little-known law of physics* to create their own gravity and their own motive force. Because the spindizzy motors worked better for higher mass, his

---

* Still undiscovered.

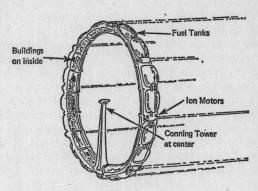

Buildings on inside

Fuel Tanks

Ion Motors

Conning Tower at center

A FLYING CITY

vehicles tended to be big. Most of the stories centered around Manhattan Island, which had been bodily uprooted from its present location and flown intact to the stars. Two of the stories involved whole worlds fitted out with spindizzies. They were even harder to land than the flying cities.

But we don't really need spindizzies or generated gravity to build flying cities.

In fact, we don't really need to fill out Heinlein's "Universe" ship. The outer hull is all we need. Visualize a ship like this:

(1) Cut a strip of Los Angeles, say, ten miles long by a mile wide.

(2) Roll it in a hoop. Buildings and streets face inward.

(3) Roof it over with glass or something stronger.

(4) Transport it to space. (Actually we'll build it in space.)

(5) Reaction motors, air and water recycling systems, and storage areas are in the basement, outward from the street level. So are the fuel tanks. Jettisoning an empty fuel tank is easy. We just cut it loose, and it falls into the universe.

(6) We're using a low-thrust, high-efficiency drive: ion jets, perhaps. The axis of the city can be kept clear. A smaller ship can rise to the axis for sightings before a course change; or we can set the control bridge atop a slender fin. A ten mile circumference makes the fin a mile and a half tall if the bridge

is at the axis; but the strain on the structure would diminish approaching the axis.

What would it be like aboard the Ring City? One gravity everywhere, except in the bridge. We may want to enlarge the bridge to accommodate a schoolroom; teaching physics would be easier in free fall.

Otherwise it would be a lot like the Generation ship. The populace would be less likely to forget their destiny, as Heinlein's people did. They can see the sky from anywhere in the city; and the only fixed stars are Sol and the target star.

It would be like living anywhere, except that great attention must be paid to environmental quality. This can be taken for granted throughout this article. The more thoroughly we control our environment, the more dangerous it is to forget it.

Inside Outside

The next step up in size is the hollow planetoid. I got my designs from a book of scientific speculation, *Islands in Space*, by Dandrige M. Cole and Donald W. Cox.

STEP ONE: Construct a giant solar mirror. Formed under zero gravity conditions, it need be nothing more than an Echo balloon sprayed with something to harden it, then cut in half and silvered on the inside. It would be fragile as a butterfly, and huge.

STEP TWO: Pick a planetoid. Ideally, we need an elongated chunk of nickel-iron, perhaps one mile in diameter and two miles long.

STEP THREE: Bore a hole down the long axis.

STEP FOUR: Charge the hole with tanks of water. Plug the openings, and weld the plugs, using the solar mirror.

STEP FIVE: Set the planetoid spinning slowly on its axis. As it spins, bathe the entire mass in the concentrated sunlight from the solar mirror. Gradually the flying iron mountain would be heated to melting all over its surface. Then the heat would creep inward, until the object is almost entirely molten.

STEP SIX: The axis would be the last part to reach melting

point. At that point the water tanks explode. The pressure blows the planetoid up into an iron balloon some ten miles in diameter and twenty miles long, if everybody has done their jobs right.

The hollow world is now ready for tenants. Except that certain things have to be moved in: air, water, soil, living things. It should be possible to set up a closed ecology. Cole and Cox suggested setting up the solar mirror at one end and using it to reflect sunlight back and forth along the long axis. We might prefer to use fusion power, if we've got it.

Naturally we spin the thing for gravity.

Living in such an inside-out world would be odd in some respects. The whole landscape is overhead. Our sky is farms and houses and so forth. If we came to space to see the stars, we'll have to go down into the basement.

We get our choice of gravity and weather. Weather is easy. We give the asteroid a slight equatorial bulge, to get a circular central lake. We shade the endpoints of the asteroid from the sun, so that it's always raining there, and the water runs downhill to the central lake. If we keep the gravity low enough, we should be able to fly with an appropriate set of muscle-powered wings; and the closer we get to the axis, the easier it becomes. (Of course, if we get too close the wax melts and the wings come apart . . .)

Macro-Life

Let's back up a bit, to the Heinlein "Universe" ship. Why do we want to land it?

If the "Universe" ship has survived long enough to reach its target star, it could probably survive indefinitely; and so can the nth-generation society it now carries. Why should their descendants live out their lives on a primitive Earthlike world? Perhaps they were born to better things.

Let the "Universe" ship become their universe, then. They can mine new materials from the asteroids of the new system, and use them to enlarge the ship when necessary, or build new ships. They can loosen the population control laws.

Change stars when convenient. Colonize space itself, and let the planets become mere way-stations. See the universe!

The concept is called *Macro-life*. Macro-life is large, powered, self-sufficient environments capable of expanding or reproducing. Put a drive on the inside-outside asteroid bubble and it becomes a Macro-life vehicle. The ring-shaped flying city can be extended indefinitely from the forward rim. Blish's spindizzy cities were a step away from being Macro-life; but they were too dependent on planet-based society.

A Macro-life vehicle would have to carry its own mining tools and chemical laboratories, and God knows what else. We'd learn what else accidentally, by losing interstellar colony ships. At best a Macro-life vehicle would never be as safe as a planet, unless it was as big as a planet, and perhaps not then. But there are other values than safety. An airplane isn't as safe as a house, but a house doesn't go anywhere. Neither does a world.

Worlds

The terraforming of worlds is the next logical step up in size. For a variety of reasons, I'm going to skip lightly over it. We know both too much and too little to talk coherently about what makes a world habitable.

But we're learning fast, and will learn faster. Our present pollution problems will end by telling us exactly how to keep a habitable environment habitable, how to keep a stable ecology stable, and how to put it all back together again after it falls apart. As usual, the universe will learn us or kill us. If we live long enough to build ships of the "Universe" type, we will know what to put inside them. We may even know how to terraform a hostile world for the convenience of human colonists, having tried our techniques on Earth itself.

Now take a giant step.

Dyson Spheres

Freeman Dyson's original argument went as follows, approximately.

No industrial society has ever reduced its need for power, except by collapsing. An intelligent optimist will expect his own society's need for power to increase geometrically, and will make his plans accordingly. According to Dyson, it will not be an impossibly long time before our own civilization needs all the power generated by our sun. Every last erg of it. We will then have to enclose the sun so as to control all of its output.

What we use to enclose the sun is problematic. Dyson was speaking of shells in the astronomical sense: solid or liquid, continuous or discontinuous, anything to interrupt the sunlight so that it can be turned into power. One move might be to convert the mass of the solar system into as many little ten-by-twenty-mile hollow iron bubbles as will fit. The smaller we subdivide the mass of a planet, the more useful surface area we get. We put all the little asteroid bubbles in circular orbits at distances of about one Earth orbit from the sun, but differing enough that they won't collide. It's a gradual process. We start by converting the existing asteroids. When we run out, we convert Mars, Jupiter, Saturn, Uranus ... and eventually, Earth.

Now, aside from the fact that our need for power increases geometrically, our population also increases geometrically. If we didn't need the power, we'd still need the room in those bubbles. Eventually we've blocked out all of the sunlight. From outside, from another star, such a system would be a great globe radiating enormous energy in the deep infrared.

What some science fiction writers have been calling a Dyson sphere is something else: a hollow spherical shell, like a ping-pong ball with a star in the middle. Mathematically at least, it is possible to build such a shell without leaving the solar system for materials. The planet Jupiter has a mass of 2×10^{30} grams, which is most of the mass of the solar system excluding the sun. Given massive transmutation of elements, we can convert Jupiter into a spherical shell 93 million miles in radius and maybe ten to twenty feet thick. If we don't have transmutation, we can still do it, with a thinner shell. There are at least ten Earthmasses of building material in the solar system, once we throw away the useless gasses.

The surface area inside a Dyson sphere is about a billion

times that of the Earth. Very few galactic civilizations in science fiction have included as many as a billion worlds. Here you'd have that much territory within walking distance, assuming you were immortal.

Naturally we would have to set up a biosphere on the inner surface. We'd also need gravity generators. The gravitational attraction inside a uniform spherical shell is zero. The net pull would come from the sun, and everything would gradually drift upward into it.

So. We spot gravity generators all over the shell, to hold down the air and the people and the buildings. "Down" is outward, toward the stars.

We can control the temperature of any locality by varying the heat-retaining properties of the shell. In fact, we may want to enlarge the shell, to give us more room or to make the permanent noonday sun look smaller. All we need do is make the shell a better insulator: foam the material, for instance. If it holds heat too well, we may want to add radiator fins to the outside.

Note that life is not necessarily pleasant in a Dyson sphere. We can't see the stars. It is always noon. We can't dig mines or basements. And if one of the gravity generators ever went out, the resulting disaster would make the end of the Earth look trivial by comparison.

But if we need a Dyson sphere, and if it can be built, we'll probably build it.

Now, Dyson's assumptions (expanding population, expanding need for power) may hold for any industrial society, human or not. If an astronomer were looking for inhabited stellar systems, he would be missing the point if he watched only the visible stars. The galaxy's most advanced civilizations may be spherical shells about the size of the Earth's orbit, radiating as much power as a Sol-type sun, but at about 10μ wavelength—in the deep infrared . . .

. . . assuming that the galaxy's most advanced civilizations are protoplasmic. But beings whose chemistry is based on molten copper, say, would want a hotter environment. They might have evolved faster, in temperatures where chemistry and biochemistry would move far faster. There might be a lot more of them than of us. And their red-hot Dyson spheres would look deceptively like red giant or supergiant stars. One wonders.

In *The Wanderer*, novelist Fritz Leiber suggested that most of the visible stars have already been surrounded by shells of worlds. We are watching old light, he suggested, light that was on its way to Earth before the industrial expansion of galactic civilization really hit its stride. Already we see some of the result: the opaque dust clouds astronomers find in the direction of the galactic core are not dust clouds, but walls of Dyson spheres blocking the stars within.

Ringworld

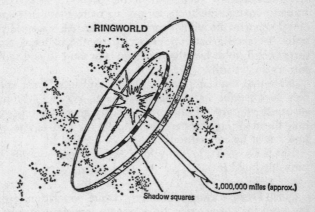

RINGWORLD

1,000,000 miles (approx.)

Shadow squares

I myself have dreamed up an intermediate step between Dyson spheres and planets. Build a ring 93 million miles in radius—one Earth orbit—which would make it 600 million miles long. If we have the mass of Jupiter to work with, and if we make it a million miles wide, we get a thickness of about a thousand meters. The Ringworld would thus be much sturdier than a Dyson sphere.

There are other advantages. We can spin it for gravity. A rotation on its axis of 770 miles/second would give the Ringworld one gravity outward. We wouldn't even have to roof it over. Put walls a thousand miles high at each rim, aimed inward at the sun, and very little of the air will leak over the edges.

Set up an inner ring of shadow squares—light orbiting structures to block out part of the sunlight—and we can have day-and-night cycles in whatever period we like. And we can see the stars, unlike the inhabitants of a Dyson sphere.

The thing is roomy enough: three million times the area of the Earth. It will be some time before anyone complains of the crowding.

As with most of these structures, our landscape is optional, a challenge to engineer and artist alike. A look at the outer surface of a Ringworld or Dyson sphere would be most instructive. Seas would show as bulges, mountains as dents. River beds and river deltas would be sculpted in; there would be no room for erosion on something as thin as a Ringworld or a Dyson sphere. Seas would be flat-bottomed—as we use only the top of a sea anyway—and small, with convoluted shorelines. Lots of beachfront. Mountains would exist only for scenery and recreation.

A large meteor would be a disaster on such a structure. A hole in the floor of the Ringworld, if not plugged, would eventually let all the air out, and the pressure differential would cause storms the size of a world, making repairs difficult.

The Ringworld concept is flexible. Consider:

(1) More than one Ringworld can circle a sun. Imagine many Ringworlds, noncoplanar, of slightly differing radii—or of widely differing radii, inhabited by very different intelligent races.

(2) We'd get seasons by bobbing the sun up and down. Actually the Ring would do the bobbing; the sun would stay put. (One Ring to a sun for this trick.)

(3) To build a Ringworld when all the planets in the system are colonized to the hilt (and, baby, we don't need a Ringworld until it's gotten that bad!) pro tem structures are needed. A structure the size of a world and the shape of a pie plate, with a huge rocket thruster underneath and a biosphere in the dish, might serve to house a planet's population while

the planet in question is being disassembled. It circles the sun at 770 miles/second, firing outward to maintain its orbit. The depopulated planet becomes two more pie plates, and we wire them in an equilateral triangle and turn off the thrusters, evacuate more planets and start building the Ringworld.

Dyson Spheres II

I pointed out earlier that gravity generators look unlikely. We may never be able to build them at all. Do we really need to assume gravity generators on a Dyson sphere? There are at least two other solutions.

We can spin the Dyson sphere. It still picks up all the energy of the sun, as planned; but the atmosphere collects around the equator, and the rest is in vacuum. We would do better to reshape the structure like a canister of movie film; it gives us greater structural strength. And we wind up with a closed Ringworld.

Or, we can live with the fact that we can't have gravity. According to the suggestion of Dan Alderson, Ph.D., we can built two concentric spherical shells, the inner shell transparent, the outer transparent or opaque, at our whim. The biosphere is between the two shells.

It would be fun. We can build anything we like within the free fall environment. Buildings would be fragile as a butterfly. Left to themselves they would drift up against the inner shell, but a heavy thread would be enough to tether them against the sun's puny gravity. The only question is, can humanity stand long periods of free fall?

Hold It A Minute

Have you reached the point of vertigo? These structures are hard to hold in your head. They're so flipping big. It might help if I tell you that, though we can't begin to build any of these things, practically anyone can handle them mathematically. Any college freshman can prove that the gravitational attraction inside a spherical shell is zero. The stresses are easy to compute (and generally too strong for anything we can make.) The mathematics of a Ringworld are those of a suspension bridge with no endpoints.

Okay, go on with whatever you were doing.

The Disc

What's bigger than a Dyson sphere? Dan Alderson, designer of the Alderson Double Dyson Sphere, now brings you the Alderson Disc. The shape is that of a phonograph record, with a sun situated in the little hole. The radius is about that of the orbit of Mars or Jupiter. Thickness: a few thousand miles.

Gravity is uniformly vertical to the surface (freshman physics again) except for edge effects. Engineers do have to worry about edge effects; so we'll build a thousand-mile wall around the inner well to keep the atmosphere from drifting into the sun. The outer edge will take care of itself.

This thing is massive. It weighs far more than the sun. We ignore problems of structural strength. Please note that we can inhabit *both* sides of the structure.

The sun will always be on the horizon, unless we bob it, which we do. (This time it is the sun that does the bobbing.) Now it is always dawn, or dusk, or night.

The Disc would be a wonderful place to stage a Gothic or a swords-and-sorcery novel. The atmosphere is right, and there are real monsters. Consider: we can occupy only a part of the

TUBE CONSTRUCTION ·
Tube — 1 mile diameter
Earth — orbit circumference approximate

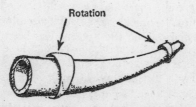

Rotation

Disc the right distance from the sun. We might as well share the Disc and the cost of its construction with aliens from hotter or colder climes. Mercurians and Venusians nearer the sun, Martians out toward the rim, aliens from other stars living wherever it suits them best. Over the tens of thousands of years, mutations and adaptations would migrate across the sparsely settled borders. If civilization should fall, things could get eerie and interesting.

Cosmic Macaroni

Pat Gunkel has designed a structure analogous to the Ringworld. Imagine a hollow strand of macaroni six hundred million miles long and not particularly thick—say a mile in diameter. Join it in a loop around the sun.

Pat calls it a *topopolis.* He points out that we could rotate the thing as in the illustration—getting gravity through centrifugal force—because of the lack of torsion effects. At six hundred million miles long and a mile wide, the curvature of the tube is negligible. We can set up a biosphere on the inner surface, with a sunlight tube down the axis and photoelectric power sources on the outside. So far, we've got something bigger than a world but smaller than a Ringworld.

But we don't have to be satisfied with one loop! We can go round and round the sun, as often as we like, as long as the strands don't touch. Pat visualizes endless loops of rotating tube, shaped like a hell of a lot of spaghetti patted roughly into a hollow sphere with a star at the center (and now we call it an *aegagropilous topopolis.*) As the madhouse civilization that built it continued to expand, the coil would reach to other stars. With the interstellar links using power supplied by the inner coils, the tube city would expand through the galaxy. Eventually our *aegagropilous galactotopopolis* would look like all the stars in the heavens had been embedded in hair.

The Megasphere

Mathematically at least, it is possible to build a really big Dyson sphere, with the heart of a galaxy at its center. There probably aren't enough planets to supply us with material.

We would have to disassemble some of the stars of the galactic arms. But we'll be able to do it by the time we need to.

We put the biosphere on the outside this time. Surface gravity is minute, but the atmospheric gradient is infinitesimal. Once again, we assume that it is possible for human beings to adapt to free fall. We live in free fall, above a surface area of tens of millions of light years, within an atmosphere that doesn't thin out for scores of light years.

Temperature control is easy: we vary the heat conductivity of the sphere to pick up and hold enough of the energy from the stars within. Though the radiating surface is great, the volume to hold heat is much greater. Industrial power would come from photoreceptors inside the shell.

Within this limitless universe of air we can build exceptionally large structures, Ringworld-sized and larger. We could even spin them for gravity. They would remain aloft for many times the lifespan of any known civilization before the gravity of the Core stars pulled them down to contact the surface.

The Megasphere would be a pleasantly poetic place to live. From a flat Earth hanging in space, one could actually reach a nearby moon via a chariot drawn by swans, and stand a good chance of finding selenites there. There would be none of this nonsense about carrying bottles of air along.

One final step to join two opposing life styles, the Macrolife tourist types and the sedentary types who prefer to restructure their home worlds.

The Ringworld rotates at 770 miles/second. Given appropriate conducting surfaces, this rotation could set up enormous magnetic effects. These could be used to control the burning of the sun, to cause it to fire off a jet of gas along the Ringworld axis of rotation. The sun becomes its own rocket. The Ringworld follows, tethered by gravity.

By the time we run out of sun, the Ring is moving through space at Bussard ramjet velocities. We continue to use the magnetic effect to pinch the interstellar gas into a fusion flame, which now becomes our sun and our motive power.

The Ringworld makes a problematical vehicle. What's it for? You can't land the damn thing anywhere. A traveling

Ringworld is not useful as a tourist vehicle; anything you want to see, you can put on the Ringworld itself ... unless it's a lovely multiple star system like Beta Lyrae; but you just can't get that close on a flying Ringworld.

A Ringworld in flight would be a bird of ill omen. It could only be fleeing some galaxy-wide disaster.

Now, galaxies do explode. We have pictures of it happening. The probable explanation is a chain reaction of novae in the galactic core. Perhaps we should be maintaining a space watch for fleeing Ringworlds ... except that we couldn't do anything about it.

We live on a world: small, immobile, vulnerable, and unprotected. But it will not be so forever.

$16,940.00

When the phone rings late at night, there is a limit to who it can be. I had three guesses as I picked it up: a wrong number (all wrong numbers are the same person), or Lois, or—I didn't bother to think his name. It isn't his anyway.

"Hello?"

"Hello," he said. "You know who this is?"

"Kelsey." It's the name he tells me. "What is it, Kelsey? You're not due for another four months."

"I need an advance. Are you sitting down?"

"I'm in bed, you son of a bitch." Reading a book, but I didn't tell him that. Better he should be off balance.

"Sorry. I just wanted you braced. I need sixteen thousand—"

"Bug off!" I slammed the phone down.

There was no point in picking up the book. He'd call again. Sometimes he waits a few minutes to make me nervous. This time the phone started ringing almost immediately, and I snatched it up in the same instant and held it to my ear without saying anything. It's a kind of bluffing game, a game I always lose.

"Kelsey again, and I'm not kidding. I need sixteen thousand, nine hundred and forty dollars. I need it by the end of the week."

"You know perfectly well I can't do that. I can't make that much money disappear without somebody noticing: Lois, the bank, the Bureau of Internal Revenue. Dammit, Kelsey, we've worked this out before."

"The best laid plans of mice and men—"

"Go to hell." Something hit me then. "That's a funny number. As long as I can't pay anyway, why not make it seventeen thousand, or twenty? Why, uh, sixteen thousand nine hundred and forty?"

"It just worked out that way." He sounded defensive.

I probed. "What way?"

"You aren't my only client."

"Client? I'm a blackmail victim! At least be honest with yourself, Kelsey."

"I am. Shall I tell you what you are?"

"No." Someone might be listening, which was the point he was trying to make. "You've got other clients, huh? Go to one of them."

"I did. It was a mistake." He hesitated, then, "Let's call him Horatio, okay? Horatio was a bank teller, long ago. He owns a hardware store now. I've known him about five years. I had to trace him myself, you understand. He embezzled some money while he was a teller."

"What did he do, die on you when the mortgage was due?" I put sarcastic sympathy in my voice.

"I wish he would. No, he waited for my usual call, which I make on April Fools Day. Not my idea; his. I call him once a year, just like you. So I called him and told him he was due, and he said he couldn't afford it any more. He got kind of brave-panicky, you know how it goes—"

"Don't I just, damn you."

"—and he said he wouldn't pay me another red cent if he had to go to prison for it. I got him to agree to meet me at a bar and grill. I hate doing that, Carson. I thought he might try to kill me."

"Occupational hazard. I may return to this subject." I had threatened to kill Kelsey before this.

He sounded dissipirited. "It won't help you. I'm careful,

Carson. I took a gun, and it was a public place, and I got there first. Besides, there are my files. If I die the cops'll go through those."

I was going to need that information, someday, maybe. But it wasn't fun to hear. "So you met him in this bar and grill. What then?"

"Well," he had the money right with him. He put it right out on the table, and I grabbed it quick because someone might be watching. Someone was, too. I saw the flashbulb go off, and by the time my eyes had stopped watering whoever it was was out the door. Ra—" He caught himself. "Horatio stopped me from getting out. He said, 'Do you know what the statute of limitations is for embezzlement?' "

"I remembered then. It was seven yeyars, and Horatio had me by the balls. Blackmail. He figures I've taken him for sixteen thousand nine hundred dollars and no cents, plus forty bucks for the guy with the camera. He wants it back or he turns me in to the police, complete with photographs."

Kelsey had never heard me laugh before and mean it. "That's hilarious. The Biter Bit bit. If you turn in your files it'll just be more evidence against you. You'll just have to fight it out in court, Kelsey. Tell 'em it's a first offense."

"I've got a better idea. I'll get the money from you."

"Nope. If I make that much money disappear, too many people would start wondering why. If they find out, I'm dead. Dead. Now I want you to remember that word, Kelsey, because it's important."

"Files, Carson. I want you to remember that word, because it's important. If I die, somebody will go through my files and then call the cops."

Well, it hadn't worked. Poor hard-luck Kelsey. "Okay, Kelsey. I'll have the money. Where can we meet?"

"No need. Just get it to me the usual way."

"Now, don't be a damn fool. I probably can't get it until Saturday, which means I'll have to get it to you Sunday. There isn't any mail Sunday."

He didn't answer for awhile. Then, "Are you thinking of killing me?"

I kept it light. "I'm always thinking of killing you, Kelsey."

"Files."

"I know. Do you want the money or don't you?"

I listened to the scared silence on the other end. Dammit, now I didn't want him scared. I was going to kill him. I'd have to find out where the files were first, and for that I'd have to have him alone, somewhere far away, for several hours. He was going to be too wary for that. I could sense it.

"Listen, there's a third way," he said suddenly. "If you move the money someone's likely to notice. If you kill me someone's sure to notice. But there's a third way."

"Let's hear it."

"Kill Horatio."

I yelped. "Kelsey, what do you think I am, Murder Incorporated? I made one mistake. One."

"You're not thinking. Carson, there is no connection between you and Horatio. None! Zilch! You can't even be suspected!"

"Um." He was right.

"You've got to do this for me, Carson. I'll never tap you for another dime . . ." He went on talking, but now I was way ahead of him. If I could get Horatio's photograph of Kelsey, I'd have Kelsey. No more payments. We'd have each other by the throats.

Poor hard-luck Horatio.

The Hole Man

One day Mars will be gone.

Andrew Lear says that it will start with violent quakes, and end hours or days later, very suddenly. He ought to know. It's all his fault.

Lear also says that it won't happen for from years to centuries. So we stay, Lear and the rest of us. We study the alien base for what it can tell us, while the center of the world we stand on is slowly eaten away. It's enough to give a man nightmares.

It was Lear who found the alien base.

We had reached Mars: fourteen of us, in the cramped bulbous life-support system of the *Percival Lowell*. We were circling in orbit, taking our time, correcting our maps and looking for anything that thirty years of Mariner probes might have missed.

We were mapping mascons, among other things. Those mass concentrations under the lunar maria were almost certainly left by good-sized asteroids, mountains of rock falling silently out of the sky until they struck with the energies of thousands of fusion bombs. Mars has been cruising through the asteroid belt for four billion years. Mars would show bigger and better mascons. They would affect our orbits.

So Andrew Lear was hard at work, watching pens twitch on graph paper as we circled Mars. A bit of machinery fell alongside the *Percival Lowell*, rotating. Within its thin shell was a weighted double lever system, deceptively simple: a Forward Mass Detector. The pens mapped its twitchings.

Over Sirbonis Palus they began mapping strange curves.

Another man might have cursed and tried to fix it. Andrew Lear thought it out, then sent the signal that would stop the free-falling widget from rotating.

It had to be rotating to map a stationery mass.

But now it was mapping simple sine waves.

Lear went running to Captain Childrey.

Running? It was more like trapeze artistry. Lear pulled himself along by handholds, kicked off from walls, braked with a hard push of hands or feet. Moving in free fall is hard work when you're in a hurry, and Lear was a forty-year-old astrophysicist, not an athlete. He was blowing hard when he reached the control bubble.

Childrey—who was an athlete—waited with a patient, slightly contemptuous smile while Lear caught his breath.

He already thought Lear was crazy. Lear's words only confirmed it. "Gravity for sending signals? Doctor Lear, will you please quit bothering me with your weird ideas. I'm busy. We all are."

This was not entirely unfair. Some of Lear's enthusiasms were peculiar. Gravity generators. Black holes. He thought we should be searching for Dyson spheres: stars completely enclosed by an artificial shell. He believed that mass and inertia were two separate things: that it should be possible to suck the inertia out of a spacecraft, say, so that it could accelerate to near lightspeed in a few minutes. He was a wide-eyed dreamer, and when he was flustered he tended to wander from the point.

"You don't understand," he told Childrey. "Gravity radiation is harder to block than electromagnetic waves. Patterned gravity waves would be easy to detect. The advanced civilizations in the galaxy may all be communicating by gravity. Some of them may even be modulating pulsars—rotating neutron stars. That's where Project Ozma went wrong: they were only looking for signals in the electromagnetic spectrum."

Childrey laughed. "Sure. Your little friends are using neutron stars to send you messages. What's what got to do with us?"

"Well, look!" Lear held up the strip of flimsy, nearly weightless paper he'd torn from the machine. "I got this over Sirbonis Palus. I think we ought to land there."

"We're landing in Mare Cimmerium, as you perfectly well know. The lander is already deployed and ready to board. Doctor Lear, we've spent four days mapping this area. It's flat. It's in a green-brown area. When spring comes next month, we'll find out whether there's life there! And everybody wants it that way except you!"

Lear was still holding the graph paper before him like a shield. "Please. Take one more circuit over Sirbonis Palus."

Childrey opted for the extra orbit. Maybe the sine waves convinced him. Maybe not. He would have liked inconveniencing the rest of us in Lear's name, to show him for a fool.

But the next pass showed a tiny circular feature in Sirbonis Palus. And Lear's mass indicator was making sine waves again.

The aliens had gone. During our first few months we always expected them back any minute. The machinery in the base was running smoothly and perfectly, as if the owners had only just stepped out.

The base was an inverted pie plate two stories high, and windowless. The air inside was breathable, like Earth's air three miles up, but with a bit more oxygen. Mars's air is far thinner, and poisonous. Clearly they were not of Mars.

The walls were thick and deeply eroded. They leaned inward against the internal pressure. The roof was somewhat thinner, just heavy enough for the pressure to support it. Both walls and roof were of fused Martian dust.

The heating system still worked—and it was also the lighting system: grids in the ceiling glowing brick red. The base was always ten degrees too warm. We didn't find the off switches for almost a week: they were behind locked panels.

The air system blew gusty winds through the base until we fiddled with the fans.

We could guess a lot about them from what they'd left behind. They must have come from a world smaller than Earth, circling a red dwarf star in close orbit. To be close enough to be warm enough, the planet would have to be locked in by tides, turning one face always to its star. The aliens must have evolved on the lighted side, in a permanent red day, with winds constantly howling over the border from the night side.

And they had no sense of privacy. The only doorways that had doors in them were airlocks. The second floor was a hexagonal metal gridwork. It would not block you off from your friends on the floor below. The bunk room was an impressive expanse of mercury-filled waterbed, wall to wall. The rooms were too small and cluttered, the furniture and machinery too close to the doorways, so that at first we were constantly bumping elbows and knees. The ceilings were an inch short of six feet high on both floors, so that we tended to walk stooped even if we were short enough to stand upright. Habit. But Lear was just tall enough to knock his head if he stood up fast, anywhere in the base.

We thought they must have been smaller than human. But their padded benches seemed human-designed in size and shape. Maybe it was their minds that were different: they didn't need psychic elbow room.

The ship had been bad enough. Now this. Within the base was instant claustrophobia. It put all of our tempers on hair triggers.

Two of us couldn't take it.

Lear and Childrey did not belong on the same planet.

With Childrey, neatness was a compulsion. He had enough for all of us. During those long months aboard *Percival Lowell*, it was Childrey who led us in calisthenics. He flatly would not let anyone skip an exercise period. We eventually gave up trying.

Well and good. The exercise kept us alive. We weren't getting the healthy daily exercise anyone gets walking around the living room in a one gravity field.

But after a month on Mars, Childrey was the only man who

still appeared fully dressed in the heat of the alien base. Some of us took it as a reproof, and maybe it was, because Lear had been the first to doff his shirt for keeps. In the mess Childrey would inspect his silverware for water spots, then line it up perfectly parallel.

On Earth, Andrew Lear's habits would have been no more than a character trait. In a hurry, he might choose mismatched socks. He might put off using the diswasher for a day or two if he were involved in something interesting. He would prefer a house that looked "lived in." God help the maid who tried to clean up his study. He'd never be able to find anything afterward.

He was a brilliant but one-sided man. Backpacking or skin diving might have changed his habits—in such pursuits you learn not to forget any least trivial thing—but they would never have tempted him. An expedition to Mars was something he simply could not turn down. A pity, because neatness is worth your life in space.

You don't leave your fly open in a pressure suit.

A month after the landing, Childrey caught Lear doing just that.

The "fly" on a pressure suit is a soft rubber tube over your male member. It leads to a bladder and there's a spring clamp on it. You open the clamp to use it. Then you close the clamp and open an outside spigot to evacuate the bladder into vacuum.

Similar designs for women involve a catheter, which is hideously uncomfortable. I presume the designers will keep trying. It seems wrong to bar half the human race from our ultimate destiny.

Lear was addicted to long walks. He loved the Martian desert scene: the hard violet sky and the soft blur of whirling orange dust, the sharp close horizon, the endless emptiness. More: he needed the room. He was spending all his working time on the alien communicator, with the ceiling too close over his head and everything else too close to his bony elbows.

He was coming back from a walk, and he met Childrey coming out. Childrey noticed that the waste spigot on Lear's suit was open, the spring broken. Lear had been out for hours. If he'd had to go, he might have bled to death through flesh ruptured by vacuum.

We never learned all that Childrey said to him out there. But Lear came in very red about the ears, muttering under his breath. He wouldn't talk to anyone.

The NASA psychologists should not have put them both on that small planet. Hindsight is wonderful, right? But Lear and Childrey were each the best choice for competence coupled to the kind of health they would need to survive the trip. There were astrophysicists as competent and as famous as Lear, but they were decades older. And Childrey had a thousand space-flight hours to his credit. He had been one of the last men on the Moon.

Individually, each of us was the best possible man. It was a damn shame.

The aliens had left the communicator going, like everything else in the base. It must have been hellishly massive, to judge by the thick support pillars slanting outward beneath it. It was a bulky tank of a thing, big enough that the roof had to bulge slightly to give it room. That gave Lear about a square meter of the only head room in the base.

Even Lear had no idea why they'd put it on the second floor. It would send through the first floor, or through the bulk of a planet. Lear learned that by trying it, once he knew enough. He beamed a dot-dash message through Mars itself to the Forward Mass Detector aboard *Lowell*.

Lear had set up a Mass Detector next to the communicator, on an extremely complex platform designed to protect it from vibration. The Detector produced waves so sharply pointed that some of us thought they could *feel* the gravity radiation coming from the communicator.

Lear was in love with the thing.

He skipped meals. When he ate he ate like a starved wolf. "There's a heavy point-mass in there," he told us, talking around a mouthful of food, two months after the landing. "The machine uses electromagnetic fields to vibrate it at high speed. Look—" He picked up a toothpaste tube of tuna spread and held it in front of him. He vibrated it rapidly. Heads turned to watch him around the zigzagged communal table in the alien mess. "I'm making gravity waves now. But they're too mushy because the tube's too big, and their amplitude is virtually zero. There's something very dense and massive in that machine, and it takes a hell of a lot of field strength to keep it there."

"What is it?" someone asked. "Neutronium? Like at the heart of a neutron star?"

Lear shook his head and took another mouthful. "That size, neutronium wouldn't be stable. I think it's a quantum black hole. I don't know how to measure its mass yet."

I said, "A quantum black hole?"

Lear nodded happily. "Luck for me. You know, I was against the Mars expedition. We could get a lot more for our money by exploring the asteroids. Among other things, we might have found if there are really quantum black holes out there. But this one's already captured!" He stood up, being careful of his head. He turned in his tray and went back to work.

I remember we stared at each other along the zigzag mess table. Then we drew lots . . . and I lost.

The day Lear left his waste spigot open, Childrey had put a restriction on him. Lear was not to leave the base without an escort.

Lear had treasured the aloneness of those walks. But it was worse than that. Childrey had given him a list of possible escorts: half a dozen men Childrey could trust to see to it that Lear did nothing dangerous to himself or others. Inevitably they were the men most thoroughly trained in space survival routines, most addicted to Childrey's own compulsive neatness, least likely to sympathize with Lear's way of living. Lear was as likely to ask Childrey himself to go walking with him.

He almost never went out any more. I knew exactly where to find him.

I stood beneath him, looking up through the gridwork floor.

He'd almost finished dismantling the protective panels around the gravity wave communicator. What showed inside looked like parts of a computer in one spot, electromagnetic coils in most places, and a square array of pushbuttons that might have been the aliens' idea of a typewriter. Lear was using a magnetic induction sensor to try to trace wiring without actually tearing off the insulation.

I called, "How you making out?"

"No good," he said. "The insulation seems to be one hundred percent perfect. Now I'm afraid to open it up. No telling how much power is running through there, if it needs shielding that good." He smiled down at me. "Let me show you something."

"What?"

He flipped a toggle above a dull grey circular plate. "This thing is a microphone. It took me awhile to find it. I am Andrew Lear, speaking to whomever may be listening." He switched it off, then ripped paper from the Mass Indicator and showed me squiggles interrupting smooth sine waves. "There. The sound of my voice in gravity radiation. It won't disappear until it's reached the edges of the universe."

"Lear, you mentioned quantum black holes back there. What's a quantum black hole?"

"Um. You know what a black hole is."

"I ought to." Lear had educated us on the subject, at length, during the months aboard *Lowell*.

When a not too massive star has used up its nuclear fuel, it collapses into a white dwarf. A heavier star—say, 1.44 times the mass of the sun and larger—can burn out its fuel, then collapse into itself until it is ten kilometers across and composed solely of neutrons packed edge to edge: the densest matter in this universe.

But a big star goes further than that. When a really massive star runs its course ... when the radiation pressure within is no longer strong enough to hold the outer layers against the star's own ferocious gravity ... then it can fall into itself entirely, until gravity is stronger than any other force, until it is compressed past the Swartzchild radius and effectively leaves the universe. What happens to it then is problematical. The Swartzchild radius is the boundary beyond which nothing can climb out of the gravity well, not even light.

The star is gone then, but the mass remains: a lightless hole in space, perhaps a hole into another universe.

"A collapsing star can leave a black hole," said Lear.

"There may be bigger black holes, whole galaxies that have fallen into themselves. But there's no other way a black hole can form, *now*."

"So?"

"There was a time when black holes of all sizes could form. That was during the Big Bang, the explosion that started the expanding universe. The forces in that blast could have compressed little local vortices of matter past the Swartzchild radius. What that left behind—the smallest ones, anyway—we call quantum black holes."

I heard a distinctive laugh behind me as Captain Childrey walked into view. The bulk of the communicator would have hidden him from Lear, and I hadn't heard him come up. He called, "Just how big a thing are you talking about? Could I pick one up and throw it at you?"

"You'd disappear into one that size," Lear said seriously. "A black hole the mass of the Earth would only be a centimeter across. No, I'm talking about things from ten-to-the-minus-fifth grams on up. There could be one at the center of the sun—"

"Eek!"

Lear was trying. He didn't like being kidded, but he didn't know how to stop it. Keeping it serious wasn't the way, but he didn't know that either. "Say, ten-to-the-seventeenth grams in mass and ten-to-the-minus-eleven centimeters across. It would be swallowing a few atoms a day."

"Well, at least you know where to find it," said Childrey. "Now all you have to do is go after it."

Lear nodded, still serious. "There could be quantum black holes in asteroids. A small asteroid could capture a quantum black hole easily enough, especially if it was charged; a black hole can hold a charge, you know—"

"Ri-ight."

"All we'd have to do is check out a small asteroid with the Mass Detector. If it masses more than it should, we push it aside and see if it leaves a black hole behind."

"You'd need little teeny eyes to see something that small. Anyway, what would you do with it?"

"You put a charge on it, if it hasn't got one already, and electromagnetic fields. You can vibrate it to make gravity;

then you manipulate it with radiation. I think I've got one in here," he said, patting the alien communicator.

"Ri-ight," said Childrey, and he went away laughing.

Within a week the whole base was referring to Lear as the Hole Man, the man with the black hole between his ears.

It hadn't sounded funny when Lear was telling me about it. The rich variety of the universe . . . but when Childrey talked about the black hole in Lear's Anything Box, it sounded hilarious.

Please note: Childrey did not misunderstand anything Lear had said. Childrey wasn't stupid. He merely thought Lear was crazy. He could not have gotten away with making fun of Lear, not among educated men, without knowing exactly what he was doing.

Meanwhile the work went on.

There were pools of Marsdust, fascinating stuff, fine enough to behave like viscous oil, and knee deep. Wading through it wasn't dangerous, but it was very hard work, and we avoided it. One day Brace waded out into the nearest of the pools and started feeling around under the dust. Hunch, he said. He came up with some eroded plastic-like containers. The aliens had used the pool as a garbage dump.

We were having little luck with chemical analysis of the base materials. They were virtually indestructible. We learned more about the chemistry of the alien visitors themselves. They had left traces of themselves on the benches and on the communal waterbed. The traces had most of the chemical components of protoplasm, but Arsvey found no sign of DNA. Not surprising, he said. There must be other giant organic molecules suitable for gene coding.

The aliens had left volumes of notes behind. The script was a mystery, of course, but we studied the photographs and diagrams. A lot of them were notes on anthropology!

The aliens had been studying Earth during the first Ice Age.

None of us were anthropologists, and that was a damn shame. We never learned if we'd found anything new. All we could do was photograph the stuff and beam it up to Lowell. One thing was sure: the aliens had left very long ago, and they had left the lighting and air systems running and the communicator sending a carrier wave.

For us? Who else?

The alternative was that the base had been switched off for some six hundred thousand years, then come back on when something detected *Lowell* approaching Mars. Lear didn't believe it. "If the power had been off in the communicator," he said, "the mass wouldn't be in there any more. The fields have to be going to hold it in place. It's smaller than an atom; it'd fall through anything solid."

So the base power system had been running for all that time. What the hell could it be? And where? We traced some cables and found that it was under the base, under several yards of Marsdust fused to lava. We didn't try to dig through that.

The source was probably geophysical: a hole deep into the core of the planet. The aliens might have wanted to dig such a hole to take core samples. Afterward they would have set up a generator to use the temperature difference between the core and the surface.

Meanwhile, Lear spent some time tracing down the power sources in the communicator. He found a way to shut off the carrier wave. Now the mass, if there was a mass, was at rest in there. It was strange to see the Forward Mass Detector pouring out straight lines instead of drastically peaked sine waves.

We were ill equipped to take advantage of these riches. We had been fitted out to explore Mars, not a bit of civilization from another star. Lear was the exception. He was in his element, with but one thing to mar his happiness.

I don't know what the final argument was about. I was engaged on another project.

The Mars lander still had fuel in it. NASA had given us plenty of fuel to hover while we looked for a landing spot. After some heated discussion, we had agreed to take the vehicle up and hover it next to the nearby dust pool on low thrust.

It worked fine. The dust rose up in a great soft cloud and went away toward the horizon, leaving the pond bottom covered with other-worldly junk. And more! Arsvey started screaming at Brace to back off. Fortunately Brace kept his head. He tilted us over to one side and took us away on a gentle curve. The backblast never touched the skeletons.

We worked out there for hours, being very finicky indeed. Here was another skill none of us would own to, but we'd read about how careful an archeologist has to be, and we did our best. Traces of water had had time to turn some of the dust to natural cement, so that some of the skeletons were fixed to the rock. But we got a couple free. We put them on stretchers and brought them back. One crumbled the instant the air came hissing into the lock. We left the other outside.

The aliens had not had the habit of taking baths. We'd set up a bathtub with very tall sides, in a room the aliens had reserved for some incomprehensible ritual. I had stripped off my pressure suit and was heading for the bathtub, very tired, hoping that nobody would be in it.

I heard the voices before I saw them.

Lear was shouting.

Childrey wasn't, but his voice was a carrying one. It carried mockery. He was standing between the supporting pillars. His hands were on his hips, his teeth gleamed white, his head was thrown back to look up at Lear.

He finished talking. For a time neither of them moved. Then Lear made a sound of disgust. He turned away and pushed one of the buttons on what might have been an alien typewriter keyboard.

Childrey looked startled. He slapped at his right thigh and brought the hand away bloody. He stared at it, then looked up at Lear. He started to ask a question.

He crumpled slowly in the low gravity. I got to him before he hit the ground. I cut his pants open and tied a handkerchief over the blood spot. It was a small puncture, but the flesh was puckered above it on a line with his groin.

Childrey tried to speak. His eyes were wide. He coughed, and there was blood in his mouth.

I guess I froze. How could I help if I couldn't tell what had happened? I saw a blood spot on his right shoulder, and I tore the shirt open and found another tiny puncture wound.

The doctor arrived.

It took Childrey an hour to die, but the doctor had given up much earlier. Between the wound in his shoulder and the wound in his thigh, Childrey's flesh had been ruptured in a narrow line that ran through one lung and his stomach and part of his intestinal tract. The autopsy showed a tiny, very neat hole drilled through the hip bones.

We looked for, and found, a hole in the floor beneath the communicator. It was the size of a pencil lead, and packed with dust.

"I made a mistake," Lear told the rest of us at the inquest. "I should never have touched that particular button. It must have switched off the fields that held the mass in place. It just dropped. Captain Childrey was underneath."

And it had gone straight through him, eating the mass of him as it went.

"No, not quite," said Lear. "I'd guess it massed about ten-to-the-fourteenth grams. That only makes it ten-to-the-minus-sixth Angstrom across, much smaller than an atom. It wouldn't have absorbed much. The damage was done to Childrey by tidal effects as it passed through him. You saw how it pulverized the material of the floor."

Not surprisingly, the subject of murder did come up.

Lear shrugged it off. "Murder with what? Childrey didn't believe there was a black hole in there at all. Neither did many of you." He smiled suddenly. "Can you imagine what the trial would be like? Imagine the prosecuting attorney trying to tell a jury what he thinks happened. First he's got to tell them what a black hole is. Then a quantum black hole. Then he's got to explain why he doesn't have the murder weapon, and where he left it, freely falling through Mars! And if he gets that far without being laughed out of court, he's still got to explain how a thing smaller than an atom could hurt anyone!"

But didn't Doctor Lear know the thing was dangerous? Could he not have guessed its enormous mass from the way it behaved?

Lear spread his hands. "Gentlemen, we're dealing with more variables than just mass. Field strength, for instance. I might have guessed its mass from the force it took to keep it there, but did any of us expect the aliens to calibrate their dials in the metric system?"

Surely there must have been safeties to keep the fields from being shut off accidentally. Lear must have bypassed them.

"Yes, I probably did, accidentally. I did quite a lot of fiddling to find out how things worked."

It got dropped there. Obviously there would be no trial. No ordinary judge or jury could be expected to understand what the attorneys would be talking about. A couple of things never did get mentioned.

For instance: Childrey's last words. I might or might not have repeated them if I'd been asked to. They were: "All right, show me! Show it to me or admit it isn't there!"

As the court was breaking up I spoke to Lear with my voice lowered. "That was probably the most unique murder weapon in history."

He whispered, "If you said that in company I could sue for slander."

"Yeah? Really? Are you going to explain to a jury what you think I implied happened?"

"No, I'll let you get away with it this time."

"Hell, you didn't get away scot free yourself. What are you going to study now? The only known black hole in the universe, and you let it drop through your fingers."

Lear frowned. "You're right. Partly right, anyway. But I knew as much about it as I was going to, the way I was going. Now . . . I stopped it vibrating in there, then took the mass of the entire setup with the Forward Mass Sensor. Now the black hole isn't in there any more. I can get the mass of the black hole by taking the mass of the communicator alone."

"Oh."

"And I can cut the machine open, see what's inside. How they controlled it. Damn it, I wish I were six years old."

"What? Why?"

"Well . . . I don't have the times straightened out. The math is chancy. Either a few years from now, or a few centuries, there's going to be a black hole between Earth and Jupiter. It'll be big enough to study. I think about forty years."

When I realized what he was implying, I didn't know whether to laugh or scream. "Lear, you can't think that something that small could absorb Mars!"

"Well, remember that it absorbs everything it comes near. A nucleus here, an electron there . . . and it's not just waiting for atoms to fall into it. Its gravity is ferocious, and it's falling back and forth through the center of the planet, sweeping up matter. The more it eats, the bigger it gets, with its volume going up as the cube of the mass. Sooner or later, yes, it'll absorb Mars. By then it'll be just less than a millimeter across—big enough to see."

"Could it happen within thirteen months?"

"Before we leave? Hmm." Lear's eyes took on a faraway look. "I don't think so. I'll have to work it out. The math is chancy . . ."

The Fourth Profession

The doorbell rang around noon on Wednesday.

I sat up in bed and—it was the oddest of hangovers. My head *didn't* spin. My sense of balance was quiveringly alert. At the same time my mind was clogged with the things I knew: facts that wouldn't relate, churning in my head.

It was like walking the high wire while simultaneously trying to solve an Agatha Christie mystery. Yet I was doing neither. I was just sitting up in bed, blinking.

I remembered the Monk, and the pills. How many pills?

The bell rang again.

Walking to the door was an eerie sensation. Most people pay no attention to their somesthetic senses. Mine were clamoring for attention, begging to be tested—by a backflip, for instance. I resisted. I don't have the muscles for doing backflips.

I couldn't remember taking any acrobatics pills.

The man outside my door was big and blond and blocky. He was holding an unfamiliar badge up to the lens of my spy-eye, in a wide hand with short, thick fingers. He had candid blue eyes, a square, honest face—a face I recognized. He'd been in the Long Spoon last night, at a single table in a corner.

147

Last night he had looked morose, introspective, like a man whose girl had left him for Mr. Wrong. A face guaranteed to get him left alone. I'd noticed him only because he wasn't drinking enough to match the face.

Today he looked patient, endlessly patient, with the patience of a dead man.

And he had a badge. I let him in.

"William Morris," he said, identifying himself. "Secret Service. Are you Edward Harley Frazer, owner of the Long Spoor Bar?"

"Part-owner."

"Yes, that's right. Sorry to bother you, Mr. Frazer. I see you keep bartender's hours." He was looking at the wrinkled pair of underpants I had on.

"Sit down," I said, waving at the chair. I badly needed to sit down myself. Standing, I couldn't think about anything but standing. My balance was all-conscious. My heels would not rest solidly on the floor. They barely touched. My weight was all on my toes; my body insisted on standing that way.

So I dropped onto the edge of the bed, but it felt like I was giving a trampoline performance. The poise, the grace, the polished ease! Hell. "What do you want from me, Mr. Morris? Doesn't the Secret Service guard the President?"

His answer sounded like rote-memory. "Among other concerns, such as counterfeiting, we do guard the President and his immediate family and the President-elect, and the Vice President if he asks us to." He paused. "We used to guard foreign dignitaries too."

That connected. "You're here about the Monk."

"Right." Morris looked down at his hands. He should have had an air of professional self-assurance to go with the badge. It wasn't there. "This is an odd case, Frazer. We took it because it used to be our job to protect foreign visitors, and because nobody else would touch it."

"So last night you were in the Long Spoon guarding a visitor from outer space."

"Just so."

"Where were you night before last?"

"Was that when he first appeared?"

"Yah," I said, remembering. "Monday night . . ."

He came in an hour after opening time. He seemed to glide, with the hem of his robe just brushing the floor. By his gait he might have been moving on wheels. His shape was wrong, in a way that made your eyes want to twist around to straighten it out.

There is something queer about the garment that gives a Monk his name. The hood is open in front, as if eyes might hide within its shadow, and the front of the robe is open too. But the loose cloth hides more than it ought to. There is too much shadow.

Once I thought the robe parted as he walked toward me. But there seemed to be nothing inside.

In the Long Spoon was utter silence. Every eye was on the Monk as he took a stool at one end of the bar, and ordered.

He looked alien, and was. But he *seemed* supernatural.

He used the oddest of drinking systems. I keep my house brands on three long shelves, more or less in order of type. The Monk moved down the top row of bottles, right to left, ordering a shot from each bottle. He took his liquor straight, at room temperature. He drank quietly, steadily, and with what seemed to be total concentration.

He spoke only to order.

He showed nothing of himself but one hand. That hand looked like a chicken's foot, but bigger, with lumpy-looking, very flexible joints, and with five toes instead of four.

At closing time the Monk was four bottles from the end of the row. He paid me in one dollar bills, and left, moving steadily, the hem of his robe just brushing the floor. I testify as an expert: he was sober. The alcohol had not affected him at all.

"Monday night," I said. "He shocked the hell out of us. Morris, what was a Monk doing in a bar in Hollywood? I thought all the Monks were in New York."

"So did we."

"Oh?"

"We didn't know he was on the West Coast until it hit the newspapers yesterday morning. That's why you didn't see more reporters yesterday. We kept them off your back. I came in last night to question you, Frazer. I changed my mind when I saw that the Monk was already here."

"Question me. Why? All I did was serve him drinks."

"Okay, let's start there. Weren't you afraid the alcohol might kill a Monk?"

"It occurred to me."

"Well?"

"I served him what he asked for. It's the Monks' own doing that nobody knows anything about Monks. We don't even know what shape they are, let alone how they're put together. If liquor does things to a Monk, it's his own lookout. Let *him* check the chemistry."

"Sounds reasonable."

"Thanks."

"It's also the reason I'm here," said Morris. "We know too little about the Monks. We didn't even know they existed until something over two years ago."

"Oh?" I'd only started reading about them a month ago.

"It wouldn't be that long, except that all the astronomers were looking in that direction already, studying a recent nova in Sagittarius. So they caught the Monk starship a little sooner; but it was already inside Pluto's orbit.

"They've been communicating with us for over a year. Two weeks ago they took up orbit around the Moon. There's only one Monk starship, and only one ground-to-orbit craft, as far as we know. The ground-to-orbit craft has been sitting in the ocean off Manhattan Island, convenient to the United Nations Building, for those same two weeks. Its crew are supposed to be all the Monks there are in the world.

"Mr. Frazer, we don't even know how your Monk got out here to the West Coast! Almost anything you could tell us would help. Did you notice anything odd about him, these last two nights?"

"Odd?" I grinned. "About a Monk?"

It took him a moment to get it, and then his answering smile was wan. "Odd for a Monk."

"Yah," I said, and tried to concentrate. It was the wrong move. Bits of fact buzzed about my skull, trying to fit themselves together.

Morris was saying, "Just talk, if you will. The Monk came back Tuesday night. About what time?"

"About four-thirty. He had a case of—pills—RNA . . ."

It was no use. I knew too many things, all at once, all unrelated. I knew the name of the Garment to Wear Among Strangers, its principle and its purpose. I knew about Monks and alcohol. I knew the names of the five primary colors, so that for a moment I was blind with the memory of the colors themselves, colors no man would ever see.

Morris was standing over me, looking worried. "What is it? What's wrong?"

"Ask me anything." My voice was high and strange and breathless with giddy laughter. "Monks have four limbs, all hands, each with a callus heel behind the fingers. I know their names, Morris. Each hand, each finger. I know how many eyes a Monk has. One. And the whole skull is an ear. There's no word for ear, but medical terms for each of the—resonating cavities—between the lobes of the brain—"

"You look dizzy. You don't sample your own wares, do you, Frazer?"

"I'm the opposite of dizzy. There's a compass in my head. I've got absolute direction. Morris, it must have been the pills."

"Pills?" Morris had small, squarish ears that couldn't possibly have come to point. But I got that impression.

"He had a sample case full of—education pills—"

"Easy now." He put a steadying hand on my shoulder. "Take it easy. Just start at the beginning, and talk. I'll make some coffee."

"Good." Coffee sounded wonderful, suddenly. "Pot's ready. Just plug it in. I fix it before I go to sleep."

Morris disappeared around the partition that marks off the kitchen alcove from the bedroom/living room in my small apartment. His voice floated back. "Start at the beginning. He came back Tuesday night."

"He came back Tuesday night," I repeated.

"Hey, your coffee's already perked. You must have plugged it in in your sleep. Keep talking."

"He started his drinking where he'd left off, four bottles

from the end of the top row. I'd have sworn he was cold sober. His voice didn't give him away . . ."

His voice didn't give him away because it was only a whisper, too low to make out. His translator spoke like a computer, putting single words together from a man's recorded voice. It spoke slowly and with care. Why not? It was speaking an alien tongue.

The Monk had had five tonight. That put him through the ryes and the bourbons and the Irish whiskeys, and several of the liqueurs. Now he was tasting the vodkas.

At that point I worked up the courage to ask him what he was doing.

He explained at length. The Monk starship was a commercial venture, a trading mission following a daisy chain of stars. He was a sampler for the group. He was mightily pleased with some of the wares he had sampled here. Probably he would order great quantities of them, to be freeze-dried for easy storage. Add alcohol and water to reconstitute.

"Then you won't be wanting to test all the vodkas," I told him. "Vodka isn't much more than water and alcohol."

He thanked me.

"The same goes for most gins, except for flavorings." I lined up four gins in front of him. One was Tanqueray. One was a Dutch gin you have to keep chilled like some liqueurs. The others were fairly ordinary products. I left him with these while I served customers.

I had expected a mob tonight. Word should have spread. *Have a drink in the Long Spoon, you'll see a Thing from Outer Space.* But the place was half empty. Louise was handling them nicely.

I was proud of Louise. As with last night, tonight she behaved as if nothing out of the ordinary was happening. The mood was contagious. I could almost hear the customers thinking: *We like our privacy when we drink. A Thing from Outer Space is entitled to the same consideration.*

It was strange to compare her present insouciance with the way her eyes had bugged at her first sight of a Monk.

The Monk finished tasting the gins. "I am concerned for

the volatile fractions," he said. "Some of your liquors will lose taste from condensation."

I told him he was probably right. And I asked, "How do you pay for your cargos?"

"With knowledge."

"That's fair. What kind of knowledge?"

The Monk reached under his robe and produced a flat sample case. He opened it. It was full of pills. There was a large glass bottle full of a couple of hundred identical pills; and these were small and pink and triangular. But most of the sample case was given over to big, round pills of all colors, individually wrapped and individually labeled in the wandering Monk script.

No two labels were alike. Some of the notations looked hellishly complex.

"These are knowledge," said the Monk.

"Ah," I said, and wondered if I was being put on. An alien can have a sense of humor, can't he? And there's no way to tell if he's lying.

"A certain complex organic molecule has much to do with memory," said the Monk. "Ribonucleic acid. It is present and active in the nervous systems of most organic beings. Wish you to learn my language?"

I nodded.

He pulled a pill loose and stripped it of its wrapping, which fluttered to the bar like a shred of cellophane. The Monk put the pill in my hand and said, "You must swallow it now, before the air ruins it, now that it is out of its wrapping."

The pill was marked like a target in red and green circles. It was big and bulky going down.

"You must be crazy," Bill Morris said wonderingly.

"It looks that way to me, too, now. But think about it. This was a Monk, an alien, an ambassador to the whole human race. He wouldn't have fed me anything dangerous, not without carefully considering all the possible consequences."

"He wouldn't, would he?"

"That's the way it seemed." I remembered about Monks and alcohol. It was a pill memory, surfacing as if I had known it all my life. It came too late . . .

"A language says things about the person who speaks it, about the way he thinks and the way he lives. Morris, the Monk language says a lot about Monks."

"Call me Bill," he said irritably.

"Okay. Take Monks and alcohol. Alcohol works on a Monk the way it works on a man, by starving his brain cells a little. But in a Monk it gets absorbed more slowly. A Monk can stay high for a week on a night's dedicated drinking.

"I knew he was sober when he left Monday night. By Tuesday night he must have been pretty high."

I sipped my coffee. Today it tasted different, and better, as if memories of some Monk staple foods had worked their way as overtones into my taste buds.

Morris said, "And you didn't know it."

"Know it? I was counting on his sense of responsibility!"

Morris shook his head in pity, except that he seemed to be grinning inside.

"We talked some more after that ... and I took some more pills."

"Why?"

"I was high on the first one."

"It made you drunk?"

"Not drunk, but I couldn't think straight. My head was full of Monk words all trying to fit themselves to meanings. I was dizzy with nonhuman images and words I couldn't pronounce."

"Just how many pills did you take?"

"I don't remember."

"Swell."

An image surfaced. "I do remember saying, 'But how about something unusual? *Really* unusual.' "

Morris was no longer amused. "You're lucky you can still talk. The chances you took, you should be a drooling idiot this morning!"

"It seemed reasonable at the time."

"You don't remember how many pills you took?"

I shook my head. Maybe the motion jarred something loose. "That bottle of little triangular pills. I know what they were. Memory erasers."

"Good God! You didn't—"

"No, no, Morris. They don't erase your whole memory. They erase pill memories. The RNA in a Monk memory pill is tagged somehow, so that the eraser pill can pick it out and break it down."

Morris gaped. Presently he said, "That's incredible. The education pills are wild enough, but *that*—You see what they must do, don't you? They hang a radical on each and every RNA molecule in each and every education pill. The active principle in the eraser pill is an enzyme for just that radical."

He saw my expression and said, "Never mind, just take my word for it. They must have had the education pills for a hundred years before they worked out the eraser principle."

"Probably. The pills must be very old."

He pounced. "How do you know that?"

"The name for the pill has only one syllable, like *fork*. There are dozens of words for kinds of pill reflexes, for swallowing the wrong pill, for side effects depending on what species is taking the pill. There's a special word for an animal training pill, and another one for a slave training pill. Morris, I think my memory is beginning to settle down."

"Good!"

"Anyway, the Monks must have been peddling pills to aliens for thousands of years. I'd guess tens of thousands."

"Just how many kinds of pill were in that case?"

I tried to remember. My head felt congested.

"I don't know if there was more than one of each kind of pill. There were four stiff flaps like the leaves of a book, and each flap had rows of little pouches with a pill in each one. The flaps were maybe sixteen pouches long by eight across. Maybe. Morris, we ought to call Louise. She probably remembers better than I do, even if she noticed less at the time."

"You mean Louise Schu the barmaid? She might at that. Or she might jar something loose in your memory."

"Right."

"Call her. Tell her we'll meet her. Where's she live, Santa Monica?"

He'd done his homework, all right.

Her phone was still ringing when Morris said, "Wait a minute. Tell her we'll meet her at the Long Spoon. And tell her we'll pay her amply for her trouble."

Then Louise answered and told me I'd jarred her out of a sound sleep, and I told her she'd be paid amply for her trouble, and she said what the hell kind of a crack was *that?*

After I hung up I asked, "Why the Long Spoon?"

"I've thought of something. I was one of the last customers out last night. I don't think you cleaned up."

"I was feeling peculiar. We cleaned up a little, I think."

"Did you empty the wastebaskets?"

"We don't usually. There's a guy who comes in in the morning and mops the floors and empties the wastebaskets and so forth. The trouble is, he's been home with flu the last couple of days. Louise and I have been going early."

"Good. Get dressed, Frazer. We'll go down to the Long Spoon and count the pieces of Monk cellophane in the wastebaskets. They shouldn't be too hard to identify. They'll tell us how many pills you took."

I noticed it while I was dressing. Morris's attitude had changed subtly. He had become proprietary. He tended to stand closer to me, as if someone might try to steal me, or as if I might try to steal away.

Imagination, maybe. But I began to wish I didn't know so much about Monks.

I stopped to empty the percolator before leaving. Habit. Every afternoon I put the percolator in the dishwasher before I leave. When I come home at three A.M. it's ready to load.

I poured out the dead coffee, took the machine apart, and stared.

The grounds in the top were fresh coffee, barely damp from steam. They hadn't been used yet.

There was another Secret Service man outside my door, a tall Midwesterner with a toothy grin. His name was George Littleton. He spoke not a word after Bill Morris introduced us, probably because I looked like I'd bite him.

I would have. My balance nagged me like a sore tooth. I couldn't forget it for an instant.

Going down in the elevator, I could feel the universe shifting around me. There seemed to be a four-dimensional map in my head, with me in the center and the rest of the universe traveling around me at various changing velocities.

The car we used was a Lincoln Continental. George drove. My map became three times as active, recording every touch of brake and accelerator.

"We're putting you on salary," said Morris, "if that's agreeable. You know more about Monks than any living man. We'll class you as a consultant and pay you a thousand dollars a day to put down all you remember about Monks."

"I'd want the right to quit whenever I think I'm mined out."

"That seems all right," said Morris. He was lying. They would keep me just as long as they felt like it. But there wasn't a thing I could do about it at the moment.

I didn't even know what made me so sure.

So I asked, "What about Louise?"

"She spent most of her time waiting on tables, as I remember. She won't know much. We'll pay her a thousand a day for a couple of days. Anyway, for today, whether she knows anything or not."

"Okay," I said, and tried to settle back.

"You're the valuable one, Frazer. You've been fantastically lucky. That Monk language pill is going to give us a terrific advantage whenever we deal with Monks. They'll have to learn about us. We'll know about them already. Frazer, what does a Monk look like under the cowl and robe?"

"Not human," I said. "They only stand upright to make us feel at ease. And there's a swelling along one side that looks like equipment under the robe, but it isn't. It's part of the digestive system. And the head is as big as a basketball, but it's half hollow."

"They're natural quadrupeds?"

"Yah. Four-footed, but climbers. The animal they evolved from lives in forests of plants that look like giant dandelions. They can throw rocks with any foot. They're still around on Center; that's the home planet. You're not writing this down."

"There's a tape recorder going."

"Really?" I'd been kidding.

"You'd better believe it. We can use anything you happen

to remember. We still don't even know how your Monk got out here to California."

My Monk, forsooth.

"They briefed me pretty quickly yesterday. Did I tell you? I was visiting my parents in Carmel when my supervisor called me yesterday morning. Ten hours later I knew just about everything anyone knows about Monks. Except you, Frazer.

"Up until yesterday we thought that every Monk on Earth was either in the United Nations Building or aboard the Monk ground-to-orbit ship.

"We've been in that ship, Frazer. Several men have been through it, all trained astronauts wearing lunar exploration suits. Six Monks landed on Earth—unless more were hiding somewhere aboard the ground-to-orbit ship. Can you think of any reason why they should do that?"

"No."

"Neither can anyone else. And there are six Monks accounted for this morning. All in New York. Your Monk went home last night."

That jarred me. "How?"

"We don't know. We're checking plane flights, silly as that sounds. Wouldn't you think a stewardess would notice a Monk on her flight? Wouldn't you think she'd go to the newspapers?"

"Sure."

"We're also checking flying saucer sightings."

I laughed. But by now that sounded logical.

"If that doesn't pan out, we'll be seriously considering teleportation. Would you—"

"That's it," I said without surprise. It had come the way a memory comes, from the back of my mind, as if it had always been there. "He gave me a teleportation pill. That's why I've got absolute direction. To teleport I've got to know where in the universe I am."

Morris got bug-eyed. "You can teleport?"

"Not from a speeding car," I said with reflexive fear. "That's death. I'd keep the velocity."

"Oh." He was edging away as if I had sprouted horns.

More memory floated up, and I said, "Humans can't teleport anyway. That pill was for another market."

Morris relaxed. "You might have said that right away."

"I only just remembered."

"Why did you take it, if it's for aliens?"

"Probably for the location talent. I don't remember. I used to get lost pretty easily. I never will again. Morris, I'd be safer on a high wire than you'd be crossing a street with the Walk sign."

"Could that have been your 'something unusual'?"

"Maybe," I said. At the same time I was somehow sure that it wasn't.

Louise was in the dirt parking lot next to the Long Spoon. She was getting out of her Mustang when we pulled up. She waved an arm like a semaphore and walked briskly toward us, already talking. "Alien creatures in the Long Spoon, forsooth!" I'd taught her that word. "Ed, I keep telling you the customers aren't human. Hello, are you Mr. Morris? I remember you. You were in last night. You had four drinks. All night."

Morris smiled. "Yes, but I tipped big. Call me Bill, okay?"

Louise Schu was a cheerful blonde, by choice, not birth. She'd been working in the Long Spoon for five years now. A few of my regulars knew my name; but they all knew hers.

Louise's deadliest enemy was the extra twenty pounds she carried as padding. She had been dieting for some decades. Two years back she had gotten serious about it and stopped cheating. She was mean for the next several months. But, clawing and scratching and half-starved every second, she had worked her way down to one hundred and twenty-five pounds. She threw a terrific celebration that night and—to hear her tell it afterward—ate her way back to one-forty-five in a single night.

Padding or not, she'd have made someone a wonderful wife. I'd thought of marrying her myself. But my marriage had been too little fun, and was too recent, and the divorce had hurt too much. And the alimony. The alimony was why I was living in a cracker box, and also the reason I couldn't afford to get married again.

While Louise was opening up, Morris bought a paper from the coin rack.

The Long Spoon was a mess. Louise and I had cleaned off the tables and collected the dirty glasses and emptied the ash trays into waste bins. But the collected glasses were still dirty and the waste bins were still full.

Morris began spreading newspaper over an area of floor.

And I stopped with my hand in my pocket.

Littleton came out from behind the bar, hefting both of the waste bins. He spilled one out onto the newspaper, then the other. He and Morris began spreading the trash apart.

My fingertips were brushing a scrap of Monk cellophane.

I'd worn these pants last night, under the apron.

Some impulse kept me from yelling out. I brought my hand out of my pocket, empty. Louise had gone to help the others sift the trash with their fingers. I joined them.

Presently Morris said, "Four. I hope that's all. We'll search the bar too."

And I thought: Five.

And I thought: I learned five new professions last night. What are the odds that I'll want to hide at least one of them?

If my judgment was bad enough to make me take a teleport pill intended for something with too many eyes, what else might I have swallowed last night?

I might be an advertising man, or a superbly trained thief, or a Palace Executioner skilled in the ways of torture. Or I might have asked for something really unpleasant, like the profession followed by Hitler or Alexander the Great.

"Nothing here," Morris said from behind the bar. Louise shrugged agreement. Morris handed the four scraps to Littleton and said, "Run these out to Douglass. Call us from there.

"We'll put them through chemical analysis," he said to Louise and me. "One of them may be real cellophane off a piece of candy. Or we might have missed one or two. For the moment, let's assume there were four."

"All right," I said.

"Does it sound right, Frazer? Should it be three, or five?"

"I don't know." As far as memory went, I really didn't.

"Four, then. We've identified two. One was a course in teleportation for aliens. The other was a language course. Right?"

"It looks that way."

"What else did he give you?"

I could feel the memories floating back there, but all scrambled together. I shook my head.

Morris looked frustrated.

"Excuse me," said Louise. "Do you drink on duty?"

"Yes," Morris said without hesitation.

And Louise and I weren't on duty. Louise mixed us three gin-and-tonics and brought them to us at one of the padded booths. Morris had opened a flattish briefcase that turned out to be part tape recorder. He said, "We won't lose anything now. Louise, let's talk about last night."

"I hope I can help."

"Just what happened in here after Ed took his first pill?"

"Mmm." Louise looked at me askance. "I don't know when he took that first pill. About one I noticed that he was acting strange. He was slow on orders. He got drinks wrong.

"I remembered that he had done that for awhile last fall, when he got his divorce—"

I felt my face go stiff. That was unexpected pain, that memory. I am far from being my own best customer; but there had been a long lost weekend about a year ago. Louise had talked me out of trying to drink and bartend too. So I had gone drinking. When it was out of my system I had gone back to tending bar.

She was saying, "Last night I thought it might be the same problem. I covered for him, said the orders twice when I had to, watched him make the drinks so he'd get them right.

"He was spending most of his time talking to the Monk. But Ed was talking English, and the Monk was making whispery noises in his throat. Remember last week, when they put the Monk speech on television? It sounded like that.

"I saw Ed take a pill from the Monk and swallow it with a glass of water."

She turned to me, touched my arm. "I thought you were crazy. I tried to stop you."

"I don't remember."

"The place was practically empty by then. Well, you laughed at me and said that the pill would teach you not to get lost! I didn't believe it. But the Monk turned on his translator gadget and said the same thing."

"I wish you'd stopped me," I said.

She looked disturbed. "I wish you hadn't said that. I took a pill myself."

I started choking. She'd caught me with a mouthful of gin and tonic.

Louise pounded my back and saved my life, maybe. She said, "You don't remember that?"

"I don't remember much of anything coherent after I took the first pill."

"Really? You didn't seem loaded. Not after I'd watched you awhile."

Morris cut in. "Louise, the pill you took. What did the Monk say it would do?"

"He never did. We were talking about me." She stopped to think. Then, baffled and amused at herself, she said, "I don't know how it happened. All of a sudden I was telling the story of my young life. To a Monk. I had the idea he was sympathetic."

"The Monk?"

"Yes, the Monk. And at some point he picked out a pill and gave it to me. He said it would help me. I believed him. I don't know why, but I believed him, and I took it."

"Any symptoms? Have you learned anything new this morning?"

She shook her head, baffled and a little truculent now. Taking that pill must have seemed sheer insanity in the cold grey light of afternoon.

"All right," said Morris. "Frazer, you took three pills. We know what two of them were. Louise, you took one, and we have no idea what it taught you." He closed his eyes a moment, then looked at me. "Frazer, if you can't remember what you took, can you remember rejecting anything? Did the Monk offer you anything—" He saw my face and cut it off.

Because that had jarred something . . .

The Monk had been speaking his own language, in that alien whisper that doesn't need to be more than a whisper because the basic sounds of the Monk language are so unambiguous, so easily distinguished, even to a human ear. *This teaches proper swimming technique. A* ——— *can reach speeds of sixteen to twenty-four* ——— *per* ——— *using these strokes. The course also teaches proper exercises . . .*

I said, "I turned down a swimming course for intelligent fish."

Louise giggled. Morris said, "You're kidding."

"I'm not. And there was something else." That swamped-in-data effect wasn't as bad as it had been at noon. Bits of data must be reaching cubbyholes in my head, linking up, finding their places.

"I was asking about the shapes of aliens. Not about Monks, because that's bad manners, especially from a race that hasn't yet proven its sentiency. I wanted to know about other aliens. So the Monk offered me three courses in unarmed combat techniques. Each one involved extensive knowledge of basic anatomy."

"You didn't take them?"

"No. What for? Like, one was a pill to tell me how to kill an armed intelligent worm, but only if I was an unarmed intelligent worm. I wasn't *that* confused."

"Frazer, there are men who would give an arm and a leg for any of those pills you turned down."

"Sure. A couple of hours ago you were telling me I was crazy to swallow an alien's education pill."

"Sorry," said Morris.

"You were the one who said they should have driven me out of my mind. Maybe they did," I said, because my hypersensitive sense of balance was still bothering the hell out of me.

But Morris's reaction bothered me worse. *Frazer could start gibbering any minute. Better pump him for all he's worth while I've got the chance.*

No, his face showed none of that. Was I going paranoid?

"Tell me more about the pills," Morris said. "It sounds like there's a lot of delayed reaction involved. How long do we have to wait before we know we've got it all?"

"He did say something . . ." I groped for it, and presently it came.

It works like a memory, the Monk had said. He'd turned off his translator and was speaking his own language, now that I could understand him. The sound of his translator had been bothering him. That was why he'd given me the pill.

But the whisper of his voice was low, and the language was new, and I'd had to listen carefully to get it all. I remembered it clearly.

The information in the pills will become part of your memory. You will not know all that you have learned until you need it. Then it will surface. Memory works by association, he'd said.

And: *There are things that cannot be taught by teachers. Always there is the difference between knowledge from school and knowledge from doing the work itself.*

"Theory and practice," I told Morris. "I know just what he meant. There's not a bartending course in the country that will teach you to leave the sugar out of an Old Fashioned during rush hour."

"What did you say?"

"It depends on the bar, of course. No posh bar would let itself get that crowded. But in an ordinary bar, anyone who orders a complicated drink during rush hour deserves what he gets. He's slowing the bartender down when it's crucial, when every second is money. So you leave the sugar out of an Old Fashioned. It's too much money."

"The guy won't come back."

"So what? He's not one of your regulars. He'd have better sense if he were."

I had to grin. Morris was shocked and horrified. I'd shown him a brand new sin. I said, "It's something every bartender ought to know about. Mind you, a bartending school is a trade school. They're teaching you to survive as a bartender. But the recipe calls for sugar, so at school you put in the sugar or you get ticked off."

Morris shook his head, tight-lipped. He said, "Then the Monk was warning you that you were getting theory, not practice."

"Just the opposite. Look at it this way, Morris—"

"Bill."

"Listen, Bill. The teleport pill can't make a human nervous

system capable of teleportation. Even my incredible balance, and it *is* incredible, won't give me the muscles to do ten quick backflips. But I do know what it *feels* like to teleport. That's what the Monk was warning me about. The pills give field training. What you have to watch out for are the reflexes. Because the pills don't change you physically."

"I hope you haven't become a trained assassin."

One must be wary of newly learned reflexes, the Monk had said.

Morris said, "Louise, we still don't know what kind of an education you got last night. Any ideas?'

"Maybe I repair time machines." She sipped her drink, eyed Morris demurely over the rim of the glass.

Morris smiled back. "I wouldn't be surprised."

The idiot. He meant it.

"If you really want to know what was in the pill," said Louise, "why not ask the Monk?" She gave Morris time to look startled, but no time to interrupt. "All we have to do is open up and wait. He didn't even get through the second shelf last night, did he, Ed?"

"No, by God, he didn't."

Louise swept an arm about her. "The place is a mess, of course. We'd never get it clean in time. Not without help. How about it, Bill? You're a government man. Could you get a team to work here in time to get this place cleaned up by five o'clock?"

"You know not what you ask. It's three fifteen now!"

Truly, the Long Spoon was a disaster area. Bars are not meant to be seen by daylight anyway. Just because our worlds had been turned upside down, and just because the Long Spoon was clearly unfit for human habitation, we had been thinking in terms of staying closed tonight. Now it was too late . . .

"Tip Top Cleaners," I remembered. "They send out a four man team with its own mops. Fifteen bucks an hour. But we'd never get them here in time."

Morris stood up abruptly. "Are they in the phone book?"

"Sure."

Morris moved.

I waited until he was in the phone booth before I asked, "Any new thoughts on what you ate last night?"

Louise looked at me closely. "You mean the pill? Why so solemn?"

"We've got to find out before Morris does."

"Why?"

"If Morris has his way," I said, "they'll classify my head Top Secret. I know too much. I'm likely to be a political prisoner the rest of my life; and so are you, if you learned the wrong things last night."

What Louise did then, I found both flattering and comforting. She turned upon the phone booth where Morris was making his call, a look of such poisonous hatred that it should have withered the man where he stood.

She believed me. She needed no kind of proof, and she was utterly on my side.

Why was I so sure? I had spent too much of today guessing at other people's thoughts. Maybe it had something to do with my third and fourth professions ...

I said, "We've got to find out what kind of pill you took. Otherwise Morris and the Secret Service will spend the rest of their lives following you around, just on the off chance that you know something useful. Like me. Only they know I know something useful. They'll be picking my brain until Hell freezes over."

Morris yelled from the phone booth. "They're coming! Forty bucks an hour, paid in advance when they get here!"

"Great!" I yelled.

"I want to call in. New York." He closed the folding door.

Louise leaned across the table. "Ed, what are we going to do?"

It was the way she said it. We were in it together, and there was a way out, and she was sure I'd find it—and she said it all in the sound of her voice, the way she leaned toward me, the pressure of her hand around my wrist. We. I felt the power and confidence rising in me; and at the same time I thought: *She couldn't do that yesterday.*

I said, "We clean this place up so we can open for business. Meanwhile you try to remember what you learned last night.

Maybe it was something harmless, like how to catch trilchies with a magnetic web."

"Tril—?"

"Space butterflies, kind of."

"Oh. But suppose he taught me how to build a faster-than-light motor?"

"We'd bloody have to keep Morris from finding out. But you didn't. The English words for going faster than light—hyperdrive, space warp—they don't have Monk translations except in math. You can't even say 'faster than light' in Monk."

"Oh."

Morris came back grinning like an idiot. "You'll never guess what the Monks want from us now."

He looked from me to Louise to me, grinning, letting the suspense grow intolerable. He said, "A giant laser cannon."

Louise gasped "What?" and I asked, "You mean a launching laser?"

"Yes, a launching laser. They want us to build it on the Moon. They'd feed our engineers pills to give them the specs and to teach them how to build it. They'd pay off in more pills."

I needed to remember something about launching lasers. And how had I known what to call it?

"They put the proposition to the United Nations," Morris was saying. "In fact, they'll be doing all of their business through the UN, to avoid charges of favoritism, they say, and to spread the knowledge as far as possible."

"But there are countries that don't belong to the UN," Louise objected.

"The Monks know that. They asked if any of those nations had space travel. None of them do, of course. And the Monks lost interest in them."

"Of course," I said, remembering. "A species that can't develop spaceflight is no better than animals."

"Huh?"

"According to a Monk."

Louise said, "But what for? Why would the Monks want a laser cannon? And on our Moon!"

"That's a little complicated," said Morris. "Do you both remember when the Monk ship first appeared, two years ago?"

"No," we answered more or less together.

Morris was shaken. "You didn't notice? It was in all the papers. Noted Astronomer Says Alien Spacecraft Approaching Earth. No?"

"No."

"For Christ's sake! I was jumping up and down. It was like when the radio astronomers discovered pulsars, remember? I was just getting out of high school."

"Pulsars?"

"Excuse me," Morris said overpolitely. "My mistake. I tend to think that everybody I meet is a science fiction fan. Pulsars are stars that give off rhythmic pulses of radio energy. The radio astronomers thought at first that they were getting signals from outer space."

Louise said, "You're a science fiction fan?"

"Absolutely. My first gun was a GyroJet rocket pistol. I bought it because I read Buck Rogers."

I said, "Buck who?" But then I couldn't keep a straight face. Morris raised his eyes to Heaven. No doubt it was there that he found the strength to go on.

"The noted astronomer was Jerome Finney. Of course he hadn't said anything about Earth. Newspapers always get that kind of thing garbled. He'd said that an object of artificial, extraterrestrial origin had entered the solar system.

"What had happened was that several months earlier, Jodrell Bank had found a new star in Sagittarius. That's the direction of the galactic core. Yes, Frazer?"

We were back to last names because I wasn't a science fiction fan. I said, "That's right. The Monks came from the galactic hub." I remembered the blazing night sky of Center. My Monk customer couldn't possibly have seen it in his lifetime. He must have been shown the vision through an education pill, for patriotic reasons, like kids are taught what the Star Spangled Banner looks like.

"All right. The astronomers were studying a nearby nova, so they caught the intruder a little sooner. It showed a strange spectrum, radically different from a nova and much more constant. It got even stranger. The light was growing brighter at the same time the spectral lines were shifting toward the red.

"It was months before anyone identified the spectrum.

"Then one Jerome Finney finally caught wise. He showed that the spectrum was the light of our own sun, drastically blue-shifted. Some kind of mirror was coming at us, moving at a hell of a clip, but slowing as it came."

"Oh!" I got it then. "That would mean a light-sail!"

"Why the big deal, Frazer? I thought you already knew."

"No. This is the first I've heard of it. I don't read the Sunday supplements."

Morris was exasperated. "But you knew enough to call the laser cannon a launching laser!"

"I just now realized why it's called that."

Morris stared at me for several seconds. Then he said, "I forgot. You got it out of the Monk language course."

"I guess so."

He got back to business. "The newspapers gave poor Finney a terrible time. You didn't see the political cartoons either? Too bad. But when the Monk ship got closer it started sending signals. It was an interstellar sailing ship, riding the sunlight on a reflecting sail, and it was coming here."

"Signals. With dots and dashes? You could do that just by tacking the sail."

"You *must* have read about it."

"Why? It's so obvious."

Morris looked unaccountably ruffled. Whatever his reasons, he let it pass. "The sail is a few molecules thick and nearly five hundred miles across when it's extended. On light pressure alone they can build up to interstellar velocities, but it takes them a long time. The acceleration isn't high.

"It took them two years to slow down to solar system velocities. They must have done a lot of braking before our telescopes found them, but even so they were going far too fast when they passed Earth's orbit. They had to go inside Mercury's orbit and come up the other side of the sun's gravity well, backing all the way, before they could get near Earth."

I said, "Sure. Interstellar speeds have to be above half the speed of light, or you can't trade competitively."

"What?"

"There are ways to get the extra edge. You don't have to depend on sunlight, not if you're launching from a civilized system. Every civilized system has a moon-based launching laser. By the time the sun is too far away to give the ship a decent push, the beam from the laser cannon is spreading just enough to give the sail a hefty acceleration without vaporizing anything."

"Naturally," said Morris, but he seemed confused.

"So that if you're heading for a strange system, you'd naturally spend most of the trip decelerating. You can't count on a strange system having a launching laser. If you know your destination is civilized, that's a different matter."

Morris nodded.

"The lovely thing about the laser cannon is that if anything goes wrong with it, there's a civilized world right there to fix it. You go sailing out to the stars with trade goods, but you leave your launching motor safely at home. Why is everybody looking at me funny?"

"Don't take it wrong," said Morris. "But how does a paunchy bartender come to know so much about flying an interstellar trading ship?"

"What?" I didn't understand him.

"Why did the Monk ship have to dive so deep into the solar system?"

"Oh, that. That's the solar wind. You get the same problem around any yellow sun. With a light-sail you can get push from the solar wind as well as from light pressure. The trouble is, the solar wind is just stripped hydrogen atoms. Light bounces from a light-sail, but the solar wind just hits the sail and sticks."

Morris nodded thoughtfully. Louise was blinking as if she had double vision.

"You can't tack against it. Tilting the sail does from nothing. To use the solar wind for braking you have to bore straight in, straight toward the sun," I explained.

Morris nodded. I saw that his eyes were as glassy as Louise's eyes.

"Oh," I said. "Damn, I must be stupid today. Morris, that was the third pill."

"Right," said Morris, still nodding, still glassy-eyed. "That must have been the unusual, *really* unusual profession you wanted. Crewman on an interstellar liner. Jesus."

And he should have sounded disgusted, but he sounded envious.

His elbows were on the table, his chin rested on his fists. It is a position that distorts the mouth, making one's expression unreadable. But I didn't like what I could read in Morris's eyes.

There was nothing left of the square and honest man I had let into my apartment at noon. Morris was a patriot now, and an altruist, and a fanatic. He must have the stars for his nation and for all mankind. Nothing must stand in his way. Least of all, me.

Reading minds again, Frazer? Maybe being captain of an interstellar liner involves having to read the minds of the crew, to be able to put down a mutiny before some idiot can take a heat point to the *mpff glip habbabub*, or however a Monk would say it; it has something to do with straining the breathing-air.

My urge to acrobatics had probably come out of the same pill. Free fall training. There was a lot in that pill.

This was the profession I should have hidden. Not the Palace Torturer, who was useless to a government grown too subtle to need such techniques; but the captain of an interstellar liner, a prize too valuable to men who have not yet reached beyond the Moon.

And I had been the last to know it. Too late, Frazer.

"Captain," I said. "Not crew."

"Pity. A crewman would know more about how to put a ship together. Frazer, how big a crew are you equipped to rule?"

"Eight and five."

"Thirteen?'

"Yes."

"Then why did you say eight and five?"

The question caught me off balance. Hadn't I . . . ? Oh. "That's the Monk numbering system. Base eight. Actually, base two, but they group the digits in threes to get base eight."

"Base two. Computer numbers."

"Are they?"

"Yes. Frazer, they must have been using computers for a long time. Aeons."

"All right." I noticed for the first time that Louise had collected our glasses and gone to make fresh drinks. Good, I could use one. She'd left her own, which was half full. Knowing she wouldn't mind, I took a swallow.

It was soda water.

With a lime in it. It had looked just like our gin and tonics. She must be back on the diet. Except that when Louise resumed a diet, she generally announced it to all and sundry . . .

Morris was still on the subject. "You use a crew of thirteen. Are they Monk or human or something else?"

"Monk," I said without having to think.

"Too bad. Are there humans in space?"

"No. A lot of two-feet, but none of them are like any of the others, and none of them are quite like us."

Louise came back with our drinks, gave them to us, and sat down without a word.

"You said earlier that a species that can't develop space flight is no better than animals."

"According to the Monks," I reminded him.

"Right. It seems a little extreme even to me, but let it pass. What about a race that develops spaceflight and then loses it?"

"It happens. There are lots of ways a space-going species can revert to animal. Atomic war. Or they just can't live with the complexity. Or they breed themselves out of food, and the world famine wrecks everything. Or waste products from the new machinery ruins the ecology."

" 'Revert to animal.' All right. What about nations? Suppose you have two nations next door, same species, but one has space flight—"

"Right. Good point, too. Morris, there are just two countries on Earth that can deal with the Monks without dealing through the United Nations. Us, and Russia. If Rhodesia or Brazil or France tried it, they'd be publicly humiliated."

"That could cause an international incident." Morris's jaw tightened heroically. "We've got ways of passing the warning along so that it won't happen."

Louise said, "There are some countries I wouldn't mind seeing it happen to."

Morris got a thoughtful look ... and I wondered if everybody would get the warning.

The cleaning team arrived then. We'd used Tip Top Cleaners before, but these four dark women were not our usual team. We had to explain in detail just what we wanted done. Not their fault. They usually clean private homes, not bars.

Morris spent some time calling New York. He must have been using a credit card; he couldn't have that much change.

"That may have stopped a minor war," he said when he got back. And we returned to the padded booth. But Louise stayed to direct the cleaning team.

The four dark women moved about us with pails and spray bottles and dry rags, chattering in Spanish, leaving shiny surfaces wherever they went. And Morris resumed his inquisition.

"What powers the ground-to-orbit ship?"

"A slow H-bomb going off in a magnetic bottle."

"Fusion?"

"Yah. The attitude jets on the main starship use fusion power too. They all link to one magnetic bottle. I don't know just how it works. You get fuel from water or ice."

"Fusion. But don't you have to separate out the deuterium and tritium?"

"What for? You melt the ice, run a current through the water, and you've got hydrogen."

"Wow," Morris said softly. "Wow."

"The launching laser works the same way," I remembered. What else did I need to remember about launching lasers? Something dreadfully important.

"Wow. Frazer, if we could build the Monks their launching laser, we could use the same techniques to build other fusion plants. Couldn't we?"

"Sure." I was in dread. My mouth was dry, my heart was pounding. I almost knew why. "What do you mean, *if*?"

"And they'd pay us to do it! It's a damn shame. We just don't have the hardware."

"What do you mean? We've got to build the launching laser!"

Morris gaped. "Frazer, what's wrong with you?"

The terror had a name now. "My God! What have you told the Monks? Morris, listen to me. You've got to see to it that the Security Council promises to build the Monks' launching laser."

"Who do you think I am, the Secretary-General? We can't build it anyway, not with just Saturn launching configurations." Morris thought I'd gone mad at last. He wanted to back away through the wall of the booth.

"They'll do it when you tell them what's at stake. And we can build a launching laser, if the whole world goes in on it. Morris, look at the good it can do! Free power from seawater! And light-sails work fine within a system."

"Sure, it's a lovely picture. We could sail out to the moons of Jupiter and Saturn. We could smelt the asteroids for their metal ores, using laser power ..." His eyes had momentarily taken on a vague, dreamy look. Now they snapped back to what Morris thought of as reality. "It's the kind of thing I daydreamed about when I was a kid. Someday we'll do it. Today—we just aren't ready."

"There are two sides to a coin," I said. "Now, I know how this is going to sound. Just remember there are reasons. Good reasons."

"Reasons? Reasons for what?"

"When a trading ship travels," I said, "it travels only from one civilized system to another. There are ways to tell whether a system has a civilization that can build a launching laser. Radio is one. The Earth puts out as much radio flux as a small star.

"When the Monks find that much radio energy coming from a nearby star, they send a trade ship. By the time the ship gets there, the planet that's putting out all the energy is generally civilized. But not so civilized that it can't use the knowledge a Monk trades for.

"Do you see that they need the launching laser? That ship out there came from a Monk colony. This far from the axis of the galaxy, the stars are too far apart. Ships launch by starlight and laser, but they brake by starlight alone, because they can't count on the target star having a launching laser. If they had to launch by starlight too, they probably wouldn't make it. A plant-and-animal cycle as small as the life support system on a Monk starship can last only so long."

"You said yourself that the Monks can't always count on the target star staying civilized."

"No, of course not. Sometimes a civilization hits the level at which it can build a launching laser, stays there just long enough to send out a mass of radio waves, then reverts to animal. That's the point. If we tell them we can't build the laser, we'll be animals to the Monks."

"Suppose we just refuse? Not *can't* but *won't*."

"That would be stupid. There are too many advantages. Controlled fusion—"

"Frazer, think about the cost." Morris looked grim. He wanted the laser. He didn't think he could get it. "Think about politicians thinking about the cost," he said. "Think about politicians thinking about explaining the cost to the taxpayers."

"Stupid," I repeated, "and inhospitable. Hospitality counts high with the Monks. You see, we're cooked either way. Either we're dumb animals, or we're guilty of a criminal breach of hospitality. And the Monk ship *still* needs more light for its light-sail than the sun can put out."

"So?"

"So the captain uses a gadget that makes the sun explode."

"The," said Morris, and "Sun," and "Explode?" He didn't know what to do. Then suddenly he burst out in great loud cheery guffaws, so that the women cleaning the Long Spoon turned with answering smiles. He'd decided not to believe me.

I reached across and gently pushed his drink into his lap.

It was two-thirds empty, but it cut his laughter off in an instant. Before he could start swearing, I said, "I am not playing games. The Monks will make our sun explode if we don't build them a launching laser. Now go call your boss and tell him so."

The women were staring at us in horror. Louise started toward us, then stopped, uncertain.

Morris sounded almost calm. "Why the drink in my lap?"

"Shock treatment. And I wanted your full attention. Are you going to call New York?"

"Not yet." Morris swallowed. He looked down once at the spreading stain on his pants, then somehow put it out of his mind. "Remember, I'd have to convince him. I don't believe it myself. Nobody and nothing would blow up a sun for a breach of hospitality!"

"No, no, Morris. They have to blow up the sun to get to the next system. It's a serious thing, refusing to build the launching laser! It could wreck the *ship!*"

"Screw the ship! What about a whole planet?"

"You're just not looking at it right—"

"Hold it. Your ship is a trading ship, isn't it? What kind of idiots would the Monks be, to exterminate one market just to get on to the next?"

"If we can't build a launching laser, we aren't a market."

"But we might be a market on the next circuit!"

"What next circuit? You don't seem to grasp the size of the Monks' marketplace. The communications gap between Center and the nearest Monk colony is about—" I stopped to transpose. "—sixty-four thousand years! By the time a ship finishes one circuit, most of the worlds she's visited have already forgotten her. And then what? The colony world that built her may have failed, or refitted the spaceport to service a different style of ship, or reverted to animal; even Monks do that. She'd have to go on to the next system for refitting.

"When you trade among the stars, *there is no repeat business.*"

"Oh," said Morris.

Louise had gotten the women back to work. With a corner of my mind I heard their giggling discussion as to whether Morris would fight, whether he could whip me, etc.

Morris asked, "How does it work? How do you make a sun go nova?"

"There's a gadget the size of a locomotive fixed to the . . . main supporting strut, I guess you'd call it. It points straight astern, and it can swing sixteen degrees or so in any direction. You turn it on when you make departure orbit. The math man works out the intensity. You beam the sun for the first year or so, and when it blows, you're just far enough away to use the push without getting burned."

"But how does it work?"

"You just turn it on. The power comes from the fusion tube that feeds the attitude jet system. —Oh, you want to know why does it make a sun explode. I don't know that. Why should I?"

"Big as a locomotive. And it makes suns explode." Morris sounded slightly hysterical. Poor bastard, he was beginning to believe me. The shock had hardly touched me, because truly I had known it since last night.

He said, "When we first saw the Monk light-sail, it was just to one side of a recent nova in Sagittarius. By any wild chance, was that star a market that didn't work out?"

"I haven't the vaguest idea."

That convinced him. If I'd been making it up, I'd have said yes. Morris stood up and walked away without a word. He stopped to pick up a bar towel on his way to the phone booth.

I went behind the bar to make a fresh drink. Cutty over ice, splash of soda; I wanted to taste the burning power of it.

Through the glass door I saw Louise getting out of her car with her arms full of packages. I poured soda over ice, squeezed a lime in it, and had it ready when she walked in.

She dumped the load on the bar top. "Irish coffee makings," she said. I held the glass out to her and she said, "No thanks, Ed. One's enough."

"Taste it."

She gave me a funny look, but she tasted what I handed her. "Soda water. Well, you caught me."

"Back on the diet?"

"Yes."

"You never said yes to that question in your life. Don't you want to tell me all the details?"

She sipped at her drink. "Details of someone else's diet are boring. I should have known that a long time ago. To work! You'll notice we've only got twenty minutes."

I opened one of her paper bags and fed the refrigerator

with cartons of whipping cream. Another bag held fresh ground coffee. The flat, square package had to be a pizza.

"Pizza. Some diet," I said.

She was setting out the glass coffee-makers. "That's for you and Bill."

I tore open the paper and bit into a pie-shaped slice. It was a deluxe, covered with everything from anchovies to salami. It was crisp and hot, and I was starving.

I snatched bites as I worked.

There aren't many bars that will keep the makings for Irish coffee handy. It's too much trouble. You need massive quantities of whipping cream and ground coffee, a refrigerator, a blender, a supply of those glass figure-eight-shaped coffee-makers, a line of hot plates, and—most expensive of all—room behind the bar for all of that. You learn to keep a line of glasses ready, which means putting the sugar in them at spare moments to save time later. Those spare moments are your smoking time, so you give that up. You learn not to wave your arms around because there are hot things that can burn you. You learn to half-whip the cream, a mere spin of the blender, because you have to do it over and over again, and if you overdo it the cream turns to butter.

There aren't many bars that will go to all that trouble. That's why it pays off. Your average Irish coffee addict will drive an extra twenty minutes to reach the Long Spoon. He'll also down the drink in about five minutes, because otherwise it gets cold. He'd have spent half an hour over a Scotch and soda.

While we were getting the coffee ready, I found time to ask, "Have you remembered anything?"

"Yes," she said.

"Tell me."

"I don't mean I know what was in the pill. Just . . . I can do things I couldn't do before. I think my way of thinking has changed. Ed, I'm worried."

"Worried?"

She got the words out in a rush. "It feels like I've been falling in love with you for a very long time. But I haven't. Why should I feel that way so suddenly?"

The bottom dropped out of my stomach. I'd had thoughts

like this ... and put them out of my mind, and when they came back I did it again. I couldn't afford to fall in love. It would cost too much. It would hurt too much.

"It's been like this all day. It scares me, Ed. Suppose I feel like this about every man? What if the Monk thought I'd make a good call girl?"

I laughed much harder than I should have. Louise was getting really angry before I was able to stop.

"Wait a minute," I said. "Are you in love with Bill Morris too?"

"No, of course not!"

"Then forget the call girl bit. He's got more money than I do. A call girl would love him more, if she loved anyone, which she wouldn't, because call girls are generally frigid."

"How do you know?" she demanded.

"I read it in a magazine."

Louise began to relax. I began to see how tense she really had been. "All right," she said, "but that means I really am in love with you."

I pushed the crisis away from us. "Why didn't you ever get married?"

"Oh ..." She was going to pass it off, but she changed her mind. "Every man I dated wanted to sleep with me. I thought that was wrong, so—"

She looked puzzled. "Why did I think that was wrong?"

"Way you were brought up."

"Yes, but ..." She trailed off.

"How do you feel about it now?"

"Well, I wouldn't sleep with anyone, but if a man was worth dating he might be worth marrying, and if he was worth marrying he'd certainly be worth sleeping with, wouldn't he? And I'd be crazy to marry someone I hadn't slept with, wouldn't I?"

"I did."

"And look how that turned out! Oh, Ed, I'm sorry. But you did bring it up."

"Yah," I said, breathing shallow.

"But I used to feel that way too. Something's changed."

We hadn't been talking fast. There had been pauses, gaps, and we had worked through them. I had had time to eat three slices of pizza. Louise had had time to wrestle with her conscience, lose, and eat one.

Only she hadn't done it. There was the pizza, staring at her, and she hadn't given it a look or a smell. For Louise, that was unusual.

Half-joking, I said, "Try this as a theory. Years ago you must have sublimated your sex urge into an urge for food. Either that or the rest of us sublimated our appetites into a sex urge, and you didn't."

"Then the pill un-sublimated me, hmm?" She looked thoughtfully at the pizza. Clearly its lure was gone. "That's what I mean. I didn't used to be able to outstare a pizza."

"Those olive eyes."

"Hypnotic, they were."

"A good call girl should be able to keep herself in shape." Immediately I regretted saying it. It wasn't funny. "Sorry," I said.

"It's all right." She picked up a tray of candles in red glass vases and moved away, depositing the candles on the small square tables. She moved with grace and beauty through the twilight of the Long Spoon, her hips swaying just enough to avoid the sharp corners of tables.

I'd hurt her. But she'd known me long enough; she must know I had foot-in-mouth disease . . .

I had seen Louise before and known that she was beautiful. But it seemed to me that she had never been beautiful with so little excuse.

She moved back by the same route, lighting the candles as she went. Finally she put the tray down, leaned across the bar and said, "I'm sorry. I can't joke about it when I don't know."

"Stop worrying, will you? Whatever the Monk fed you, he was trying to help you."

"I love you."

"What?"

"I love you."

"Okay. I love you too." I use those words so seldom that they clog in my throat, as if I'm lying, even when it's the truth. "Listen, I want to marry you. Don't shake your head. I want to marry you."

Our voices had dropped to whispers. In a tormented whisper, then, she said, "Not until I find out what I *do*, what was in the *pill*. Ed, I can't trust myself until then!"

"Me too," I said with great reluctance. "But we can't wait. We don't have time."

"What?"

"That's right, you weren't in earshot. Sometime between three and ten years from now, the Monks may blow up our sun."

Louise said nothing. Her forehead winkled.

"It depends on how much time they spend trading. If we can't build them the launching laser, we can still con them into waiting for awhile. Monk expeditions have waited as long as—"

"Good Lord. You mean it. Is that what you and Bill were fighting over?"

"Yah."

Louise shuddered. Even in the dimness I saw how pale she had become. And she said a strange thing.

She said, "All right, I'll marry you."

"Good," I said. But I was suddenly shaking. Married. Again. Me. Louise stepped up and put her hands on my shoulders, and I kissed her.

I'd been wanting to do that for ... five years? She fitted wonderfully into my arms. Her hands closed hard on the muscles of my shoulders, massaging. The tension went out of me, drained away somewhere. Married. Us. At least we could have three to ten years.

"Morris," I said.

She drew back a little. "He can't hold you. You haven't done anything. Oh, I *wish* I knew what was in that pill I took! Suppose I'm the trained assassin?"

"Suppose I am? We'll have to be careful of each other."

"Oh, we know all about you. You're a starship commander, an alien teleport and a translator for Monks."

"And one thing more. There was a fourth profession. I took four pills last night, not three."

"Oh? Why didn't you tell Bill?"

"Are you kidding? Dizzy as I was last night, I probably took a course in how to lead a successful revolution. God help me if Morris found *that* out."

She smiled. "Do you really think that was what it was?"

"No, of course not."

"Why did we do it? Why did we swallow those pills? We should have known better."

"Maybe the Monk took a pill himself. Maybe there's a pill that teaches a Monk how to look trustworthy to a generalized alien."

"I did trust him," said Louise. "I remember. He seemed so sympathetic. Would he really blow up our sun?"

"He really would."

"That fourth pill. Maybe it taught you a way to stop him."

"Let's see. We know I took a linguistics course, a course in teleportation for martians, and a course in how to fly a light-sail ship. On that basis . . . I probably changed my mind and took a karate course for worms."

"It wouldn't hurt you, at least. Relax. . . . Ed, if you remember taking the pills, why don't you remember what was in them?"

"But I don't. I don't remember anything."

"How do you know you took four, then?"

"Here." I reached in my pocket and pulled out the scrap of Monk cellophane. And knew immediately that there was something in it. Something hard and round.

We were staring at it when Morris came back.

"I must have cleverly put it in my pocket," I told them. "Sometime last night, when I was feeling sneaky enough to steal from a Monk."

Morris turned the pill like a precious jewel in his fingers. Pale blue it was, marked on one side with a burnt orange triangle. "I don't know whether to get it analyzed or take it myself, now. We need a miracle. Maybe this will tell us—"

"Forget it. I wasn't clever enough to remember how fast a Monk pill deteriorates. The wrapping's torn. That pill has been bad for at least twelve hours."

Morris said a dirty thing.

"Analyze it," I said. "You'll find RNA, and you may even be able to tell what the Monks use as a matrix. Most of the memories are probably intact. But don't swallow the damn thing. It'll scramble your brains. All it takes is a few random changes in a tiny percentage of the RNA."

"We don't have time to send it to Douglass tonight. Can we put it in the freezer?"

"Good. Give it here."

I dropped the pill in a sandwich-size plastic Baggy, sucked the air out the top, tied the end, and dropped it in the freezer. Vacuum and cold would help preserve the thing. It was something I should have done last night.

"So much for miracles," Morris said bitterly. "Let's get down to business. We'll have several men outside the place tonight, and a few more in here. You won't know who they are, but go ahead and guess if you like. A lot of your customers will be turned away tonight. They'll be told to watch the newspapers if they want to know why. I hope it won't cost you too much business."

"It may make our fortune. We'll be famous. Were you maybe doing the same thing last night?"

"Yes. We didn't want the place too crowded. The Monks might not like autograph hounds."

"So that's why the place was half empty."

Morris looked at his watch. "Opening time. Are we ready?"

"Take a seat at the bar. And look nonchalant, dammit."

Louise went to turn on the lights.

Morris took a seat to one side of the middle. One big square hand was closed very tightly on the bar edge. "Another gin and tonic. Weak. After that one, leave out the gin."

"Right."

"Nonchalant. Why should I be nonchalant? Frazer, I had to tell the President of the United States of America that the end of the world is coming unless he does something. I had to talk to him myself!"

"Did he buy it?"

"I hope so. He was so goddam calm and reassuring, I wanted to scream at him. God, Frazer, what if we can't build the laser? What if we try and fail?"

I gave him a very old and classic answer. "Stupidity is always a capital crime."

He screamed in my face. "Damn you and your supercilious

attitude and your murdering monsters too!" The next second he was ice-water calm. "Never mind, Frazer. You're thinking like a starship captain."

"I'm what?"

"A starship captain has to be able to make a sun go nova to save the ship. You can't help it. It was in the pill."

Damn, he was right. I could *feel* that he was right. The pill had warped my way of thinking. Blowing up the sun that warms another race *had* to be immoral. Didn't it?

I couldn't trust my own sense of right and wrong!

Four men came in and took one of the bigger tables. Morris's men? No. Real estate men, here to do business.

"Something's been bothering me," said Morris. He grimaced. "Among all the things that have been ruining my composure, such as the impending end of the world, there was one thing that kept nagging at me."

I set his gin-and-tonic in front of him. He tasted it and said, "Fine. And I finally realized what it was, waiting there in the phone booth for a chain of human snails to put the President on. Frazer, are you a college man?"

"No. Webster High."

"See, you don't really talk like a bartender. You use big words."

"I do?"

"Sometimes. And you talked about 'suns exploding,' but you knew what I meant when I said 'nova.' You talked about 'H-bomb power,' but you knew what fusion was."

"Sure."

"I got the possibly silly impression that you were learning the words the instant I said them. *Parlez-vous français?*"

"No. I don't speak any foreign languages."

"None at all?"

"Nope. What do you think they teach at Webster High?"

"*Je parle la langue un peu*, Frazer. *Et tu?*"

"*Merde de cochon!* Morris, *je vous dit*—oops."

He didn't give me a chance to think it over. He said, "What's *fanac*?"

My head had that clogged feeling again. I said, "Might be anything. Putting out a zine, writing to the lettercol, helping put on a Con—Morris, what *is* this?"

"That language course was more extensive than we thought."

"Sure as hell, it was. I just remembered. Those women on the cleaning team were speaking Spanish, but I understood them."

"Spanish, French, Monkish, technical languages, even Fannish. What you got was a generalized course in how to understand languages the instant you hear them. I don't see how it could work without telepathy."

"Reading minds? Maybe." Several times today, it had felt like I was guessing with too much certainty at somebody's private thoughts.

"Can you read *my* mind?"

"That's not quite it. I get the feel of *how* you think, not *what* you're thinking. Morris, I don't like the idea of being a political prisoner."

"Well, we can talk that over later." *When my bargaining position is better,* Morris meant. *When I don't need the bartender's good will to con the Monk.* "What's important is that you might be able to read a Monk's mind. That could be crucial."

"And maybe he can read mine. And yours."

I let Morris sweat over that one while I set drinks on Louise's tray. Already there were customers at four tables. The Long Spoon was filling rapidly; and only two of them were Secret Service.

Morris said, "Any ideas on what Louise Schu ate last night? We've got your professions pretty well pegged down. Finally."

"I've got an idea. It's kind of vague." I looked around. Louise was taking more orders. "Sheer guesswork, in fact. Will you keep it to yourself for awhile?"

"Don't tell Louise? Sure—for awhile."

I made four drinks, and Louise took them away. I told Morris, "I have a profession in mind. It doesn't have a simple one or two word name, like teleport or starship captain or translator. There's no reason why it should, is there? We're dealing with aliens."

Morris sipped at his drink. Waiting.

"Being a woman," I said, "can be a profession, in a way

that being a man can never be. The word is *housewife*, but it doesn't cover all of it. Not nearly."

"Housewife. You're putting me on."

"No. You wouldn't notice the change. You never saw her before last night."

"Just what kind of change have you got in mind? Aside from the fact that she's beautiful, which I did notice."

"Yes, she is, Morris. But last night she was twenty pounds overweight. Do you think she lost it all this morning?"

"She was too heavy. Pretty, but also pretty well padded." Morris turned to look over his shoulder, casually turned back. "Damn. She's still well padded. Why didn't I notice before?"

"There's another thing.—By the way. Have some pizza."

"Thanks." He bit into a slice. "Good, it's still hot. Well?"

"She's been staring at that pizza for half an hour. She bought it. But she hasn't tasted it. She couldn't possibly have done that yesterday."

"She may have had a big breakfast."

"Yah." I knew she hadn't. She'd eaten diet food. For years she'd kept a growing collection of diet food, but she'd never actively tried to survive on it before. But how could I make such a claim to Morris? I'd never even been in Louise's apartment.

"Anything else?"

"She's gotten good at nonverbal communication. It's a very womanly skill. She can say things just by the tone of her voice or the way she leans on an elbow or—"

"But if mind reading is one of your new skills . . ."

"Damn. Well . . . it used to make Louise nervous if someone touched her. And she never touched anyone else." I felt myself flushing. I don't talk easily of personal things.

Morris radiated skepticism. "It all sounds very subjective. In fact, it sounds like you're making yourself believe it. Frazer, why would Louise Schu want such a capsule course? Because you haven't described a housewife at all. You've described a woman looking to persuade a man to marry her." He saw my face change. "What's wrong?"

"Ten minutes ago we decided to get married."

"Congratulations," Morris said, and waited.

"All right, you win. Until ten minutes ago we'd never even

kissed. I'd never made a pass, or vice versa. No, damn it, I don't believe it! I know she loves me; I ought to!"

"I don't deny it," Morris said quietly. "That would be why she took the pill. It must have been strong stuff, too, Frazer. We looked up some of your history. You're marriage-shy."

It was true enough. I said, "If she loved me before, I never knew it. I wonder how a Monk could know."

"How would he know about such a skill at all? Why would he have the pill on him? Come on, Frazer, you're the Monk expert!"

"He'd have to learn from human beings. Maybe by interviews, maybe by—well, the Monks can map an alien memory into a computer space, then interview that. They may have done that with some of your diplomats."

"Oh, great."

Louise appeared with an order. I made the drinks and set them on her tray. She winked and walked away, swaying deliciously, followed by many eyes.

"Morris. Most of your diplomats, the ones who deal with the Monks; they're men, aren't they?"

"Most of them. Why?"

"Just a thought."

It was a difficult thought, hard to grasp. It was only that the changes in Louise had been all to the good from a man's point of view. The Monks must have interviewed many men. Well, why not? It would make her more valuable to the man she caught—or to the lucky man who caught her—

"Got it."

Morris looked up quickly. "Well?"

"Falling in love with me was part of her pill learning. A set. They made a guinea pig of her."

"I wondered what she saw in you." Morris's grin faded. "You're serious. Frazer, that still doesn't answer—"

"It's a slave indoctrination course. It makes a woman love the first man she sees, permanently, and it trains her to be valuable to him. The Monks were going to make them in quantity and sell them to men."

Morris thought it over. Presently he said, "That's awful. What'll we do?"

"Well, we can't tell her she's been made into a domestic

slave! Morris, I'll try to get a memory eraser pill. If I can't
... I'll marry her, I guess. Don't look at me that way," I said,
low and fierce. "I didn't do it. And I can't desert her now!"

"I know. It's just—oh, put gin in the next one."

"Don't look now," I said.

In the glass of the door there was darkness and motion. A
hooded shape, shadow-on-shadow, supernatural, a human sil-
houette twisted out of true ...

He came gliding in with the hem of his robe just brushing
the floor. Nothing was to be seen of him but his flowing gray
robe, the darkness in the hood and the shadow where his robe
parted. The real estate men broke off their talk of land and
stared, popeyed, and one of them reached for his heart attack
pills.

The Monk drifted toward me like a vengeful ghost. He
took the stool we had saved him at one end of the bar.

It wasn't the same Monk.

In all respects he matched the Monk who had been here
these last two nights. Louise and Morris must have been
fooled completely. But it wasn't the same Monk.

"Good evening," I said.

He gave an equivalent greeting in the whispered Monk language. His translator was half on, translating my words into a
Monk whisper, but letting his own speech alone. He said, "I
believe we should begin with the Rock and Rye."

I turned to pour. The small of my back itched with danger.

When I turned back with the shot glass in my hand, he was
holding a fist-sized tool that must have come out of his robe.
It looked like a flattened softball, grooved deeply for five
Monk claws, with two parallel tubes poking out in my direction. Lenses glinted in the ends of the tubes.

"Do you know this tool? It is a ———" and he named it.

I knew the name. It was a beaming tool, a multi-frequency
laser. One tube locked on the target; thereafter the aim was
maintained by tiny flywheels in the body of the device.

Morris had seen it. He didn't recognize it, and he didn't
know what to do about it, and I had no way to signal him.

"I know that tool," I confirmed.

"You must take two of these pills." The Monk had them

ready in another hand. They were small and pink and triangular. He said, "I must be convinced that you have taken them. Otherwise you must take more than two. An overdose may affect your natural memory. Come closer."

I came closer. Every man and woman in the Long Spoon was staring at us, and each was afraid to move. Any kind of signal would have trained four guns on the Monk. And I'd be fried dead by a narrow beam of X-rays.

The Monk reached out with a third hand/foot/claw. He closed the fingers/toes around my throat, not hard enough to strangle me, but hard enough.

Morris was cursing silently, helplessly. I could feel the agony in his soul.

The Monk whispered, "You know of the trigger mechanism. If my hand should relax now, the device will fire. Its target is yourself. If you can prevent four government agents from attacking me, you should do so."

I made a palm-up gesture toward Morris. *Don't do anything.* He caught it and nodded very slightly without looking at me.

"You can read minds," I said.

"Yes," said the Monk—and I knew instantly what he was hiding. He could read everybody's mind, except mine.

So much for Morris's little games of deceit. But the Monk could not read my mind, and I could see into his own soul.

And, reading his alien soul, I saw that I would die if I did not swallow the pills.

I placed the pink pills on my tongue, one at a time, and swallowed them dry. They went down hard. Morris watched it happen and could do nothing. The Monk felt them going down my throat, little lumps moving past his finger.

And when the pills had passed across the Monk's finger, I worked a miracle.

"Your pill-induced memories and skills will be gone within two hours," said the Monk. He picked up the shot glass of Rock and Rye and moved it into his hood. When it reappeared it was half empty.

I asked. "Why have you robbed me of my knowledge?"

"You never paid for it."

"But it was freely given."

"It was given by one who had no right," said the Monk. He was thinking about leaving. I had to do something. I knew now, because I had reasoned it out with great care, that the Monk was involved in an evil enterprise. But he must stay to hear me or I could not convince him.

Even then, it wouldn't be easy. He was a Monk crewman. His ethical attitudes had entered his brain through an RNA pill, along with his professional skills.

"You have spoken of rights," I said. In Monk. "Let us discuss rights." The whispery words buzzed oddly in my throat; they tickled; but my ears told me they were coming out right.

The Monk was startled. "I was told that you had been taught our speech, but not that you could speak it."

"Were you told what pill I was given?"

"A language pill. I had not known that he carried one in his case."

"He did not finish his tasting of the alcohols of Earth. Will you have another drink?"

I felt him guess at my motives, and guess wrong. He thought I was taking advantage of his curiosity to sell him my wares for cash. And what had he to fear from me? Whatever mental powers I had learned from Monk pills, they would be gone in two hours.

I set a shot glass before him. I asked him, "How do you feel about launching lasers?"

The discussion became highly technical. "Let us take a special case," I remember saying. "Suppose a culture has been capable of starflight for some sixty-fours of years—or even for eights of times that long. Then an asteroid slams into a major ocean, precipitates an ice age . . ." It had happened once, and well he knew it. "A natural disaster can't spell the difference between sentience and nonsentience, can it? Not unless it affects brain tissue directly."

At first it was his curiosity that held him. Later it was me. He couldn't tear himself loose. He never thought of it. He was a sailship crewman, and he was cold sober, and he argued with the frenzy of an evangelist.

"Then take the general case," I remember saying. "A world

that cannot build a launching laser is a world of animals, yes? And Monks themselves can revert to animals."

Yes, he knew that.

"Then build your own launching laser. If you cannot, then your ship is captained and crewed by animals."

At the end I was doing all the talking. All in the whispery Monk tongue, whose sounds are so easily distinguished that even I, warping a human throat to my will, need only whisper. It was a good thing. I seemed to have been eating used razor blades.

Morris guessed right. He did not interfere. I could tell him nothing, not if I had had the power, not by word or gesture or mental contact. The Monk would read Morris's mind. But Morris sat quietly drinking his tonic-and-tonics, waiting for something to happen. While I argued in whispers with the Monk.

"But the ship!" he whispered. "What of the ship?" His agony was mine; for the ship must be protected . . .

At one fifteen the Monk was halfway across the bottom row of bottles. He slid from the stool, paid for his drinks in one dollar bills, and drifted to the door and out.

All he needed was a scythe and hour glass, I thought, watching him go. And what I needed was a long morning's sleep. And I wasn't going to get it.

"Be sure nobody stops him," I told Morris.

"Nobody will. But he'll be followed."

"No point. The Garment to Wear Among Strangers is a lot of things. It's bracing; it helps the Monk hold human shape. It's a shield and an air filter. And it's a cloak of invisibility."

"Oh?"

"I'll tell you about it if I have time. That's how he got out here, probably. One of the crewmen divided, and then one stayed and one walked. He had two weeks."

Morris stood up and tore off his sport jacket. His shirt was wet through. He said, "What about a stomach pump for you?"

"No good. Most of the RNA-enzyme must be in my blood by now. You'll be better off if you spend your time getting down everything I can remember about Monks, while I can remember anything at all. It'll be nine or ten hours before everything goes." Which was a flat-out lie, of course.

"Okay. Let me get the dictaphone going again."

"It'll cost you money."

Morris suddenly had a hard look. "Oh? How much?"

I'd thought about that most carefully. "One hundred thousand dollars. And if you're thinking of arguing me down, remember whose time we're wasting."

"I wasn't." He was, but he'd changed his mind.

"Good. We'll transfer the money now, while I can still read your mind."

"All right."

He offered to make room for me in the booth, but I declined. The glass wouldn't stop me from reading Morris's soul.

He came out silent; for there was something he was afraid to know. Then: "What about the Monks? What about our sun?"

"I talked that one around. That's why I don't want him molested. He'll convince others."

"Talked him around? How?"

"It wasn't easy." And suddenly I would have given my soul to sleep. "The profession pill put it in his genes; he must protect the ship. It's in me too. I know how strong it is."

"Then—"

"Don't be an ass, Morris. The ship's perfectly safe where it is, in orbit around the Moon. A sailship's only in danger when it's between stars, far from help."

"Oh."

"Not that that convinced him. It only let him consider the ethics of the situation rationally."

"Suppose someone else unconvinces him?"

"It could happen. That's why we'd better build the launching laser."

Morris nodded unhappily.

The next twelve hours were rough.

In the first four hours I gave them everything I could remember about the Monk teleport system, Monk technology, Monk family life, Monk ethics, relations between Monks and aliens, details on aliens, directions of various inhabited and uninhabited worlds ... everything. Morris and the Secret Service men who had been posing as customers sat around me like boys around a campfire, listening to stories. But Louise made us fresh coffee, then went to sleep in one of the booths.

Then I let myself slack off.

By nine in the morning I was flat on my back, staring at the ceiling, dictating a random useless bit of information every thirty seconds or so. By eleven there was a great black pool of lukewarm coffee inside me, my eyes ached marginally more than the rest of me, and I was producing nothing.

I was convincing, and I knew it.

But Morris wouldn't let it go at that. He believed me. I felt him believing me. But he was going through the routine anyway, because it couldn't hurt. If I was useless to him, if I knew nothing, there was no point in playing soft. What could he lose?

He accused me of making everything up. He accused me of faking the pills. He made me sit up, and damn near caught me that way. He used obscure words and phrases from mathematics and Latin and Fan vocabulary. He got nowhere. There wasn't any way to trick me.

At two in the afternoon he had someone drive me home.

Every muscle in me ached; but I had to fight to maintain my exhausted slump. Else my hindbrain would have lifted me onto my toes and poised me against a possible shift in artificial gravity. The strain was double, and it hurt. It had hurt for hours, sitting with my shoulders hunched and my head hanging. But now, if Morris saw me walking like a trampoline performer ...

Morris's man got me to my room and left me.

I woke in darkness and sensed someone in my room. Someone who meant me no harm. In fact, Louise. I went back to sleep.

I woke again at dawn. Louise was in my easy chair, her feet propped on a corner of the bed. Her eyes were open. She said, "Breakfast?"

I said, "Yah. There isn't much in the fridge."

"I brought things."

"All right." I closed my eyes.

Five minutes later I decided I was all slept out. I got up and went to see how she was doing.

There was bacon frying, there was bread already buttered for toasting in the Toast-R-Oven, there was a pan hot for eggs, and the eggs scrambled in a bowl. Louise was filling the percolator.

"Give that here a minute," I said. It only had water in it. I held the pot in my hands, closed my eyes and tried to remember . . .

Ah.

I knew I'd done it right even before the heat touched my hands. The pot held hot, fragrant coffee.

"We were wrong about the first pill," I told Louise. She was looking at me very curiously. "What happened that second night was this. The Monk had a translator gadget, but he wasn't too happy with it. It kept screaming in his ear. Screaming English.

"He could turn off the part that was shouting English at me, and it would still whisper a Monk translation of what I was saying. But first he had to teach me the Monk language. He didn't have a pill to do that. He didn't have a generalized language-learning course either, if there is one, which I doubt.

"He was pretty drunk, but he found something that would serve. The profession it taught me was something like yours. I mean, it's an old one, and it doesn't have a one-or-two-word name. But if it did, the word would be *prophet*."

"Prophet," said Louise. "Prophet?" She was doing a remarkable thing. She was listening with all her concentration, and scrambling eggs at the same time.

"Or disciple. Maybe *apostle* comes closer. Anyway, it included the Gift of Tongues, which was what the Monk was after. But it included other talents too."

"Like turning cold water into hot coffee?"

"Miracles, right. I used the same talent to make the little pink amnesia pills disappear before they hit my stomach. But an apostle's major talent is persuasion.

"Last night I convinced a Monk crewman that blowing up suns is an evil thing.

"Morris is afraid that someone might convert him back. I

don't think that's possible. The mind reading talent that goes with the prophet pill goes deeper than just reading minds. I read souls. The Monk is my apostle. Maybe he'll convince the whole crew that I'm right.

"Or he may just curse the *hachiroph shisp*, the little old nova maker. Which is what I intend to do."

"Curse it?"

"Do you think I'm kidding or something?"

"Oh, no." She poured our coffee. "Will that stop it working?"

"Yes."

"Good," said Louise. And I felt the power of her own faith, her faith in me. It gave her the serenity of an idealized nun.

When she turned back to serve the eggs, I dropped a pink triangular pill in her coffee.

She finished setting breakfast and we sat down. Louise said, "Then that's it. It's all over."

"All over." I swallowed some orange juice. Wonderful, what fourteen hours' sleep will do for a man's appetite. "All over. I can go back to my fourth profession, the only one that counts."

She looked up quickly.

"Bartender. First, last, and foremost, I'm a bartender. You're going to marry a bartender."

"Good," she said, relaxing.

In two hours or so the slave sets would be gone from her mind. She would be herself again: free, independent, unable to diet, and somewhat shy.

But the pink pill would not destroy real memories. Two hours from now, Louise would still know that I loved her; and perhaps she would marry me after all.

I said, "We'll have to hire an assistant. And raise our prices. They'll be fighting their way in when the story gets out."

Louise had pursued her own thoughts. "Bill Morris looked awful when I left. You ought to tell him he can stop worrying."

"Oh, no. I *want* him scared. Morris has got to talk the rest of the world into building a launching laser, instead of just throwing bombs at the Monk ship. And we need the launching laser."

"Mmm! That's good coffee. Why do we need a launching laser?"

"To get to the stars."

"That's Morris's bag. You're a bartender, remember? The fourth profession."

I shook my head. "You and Morris. You don't see how *big* the Monk marketplace is, or how thin the Monks are scattered. How many novas have you seen in your lifetime?

"Damn few," I said. "There are damn few trading ships in a godawful lot of sky. There are things out there besides Monks. Things the Monks are afraid of, and probably others they don't know about.

"Things so dangerous that the only protection is to be somewhere else, circling some other star, when it happens here! The Monk drive is our lifeline and our immortality. It would be cheap at any price—"

"Your eyes are glowing," she breathed. She looked half hypnotized, and utterly convinced. And I knew that for the rest of my life, I would have to keep a tight rein on my tendency to preach.